The Illustrated Guide to Video Formats

Written and illustrated by
Ashley Blewer

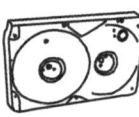

The Illustrated Guide to Video Formats

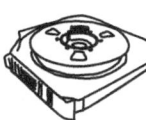
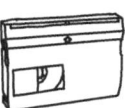

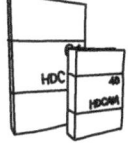
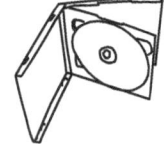

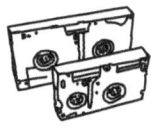

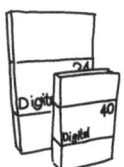

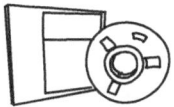

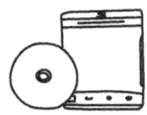

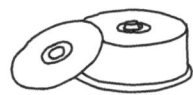

Written and illustrated by
Ashley Blewer

Published by Archives of Tomorrow LLC, Philadelphia, Pennsylvania

https://archivesoftomorrow.com

Cover design by Ashley Blewer
Illustrations by Ashley Blewer

ISBN: 978-1-958543-00-9
Library of Congress Control Number: 2022910037

First Edition

This one's for the archivists

Table of Contents

Introduction

This book provides an introductory overview to thirty-six historically significant video formats developed between 1956–2006. This collection begins with the first major video reel format, 2" Quadruplex video (invented in 1956) and concludes with the Blu-ray disc (introduced in 2006).

AKA: Any other notable names that this format was "also known as"

Format: Whether the format was primarily analog or digital

Developed by: The primary developer, manufacturer, or patent holder(s) for the format

Era: The approximate era of production (acknowledging that end dates are particularly fuzzy, as some formats continued to be used long past their production expiration date)

Capacity: The *maximum* capacity duration that the format was able to hold

Size: The physical dimension(s) most common for the format, measured in inches (for reel diameters) or centimeters (for cartridges or other rectangular objects)

There are at least twice as many formats that went into production during the video era, and not all are accounted for in this guide. Furthermore, some formats have been gathered into families (e.g. Betacam, DVCPro, U-matic) or some family pairs (e.g. V-Cord and V-Cord II) for the sake of brevity.

This book does its best to give a high-level summary of formats as they might be discovered. There are so many more interesting technical aspects of each format that did not fit into this book. To name just a few: tape width, thickness, material, audio channels, bands, bandwidths, sampling frequencies, color spaces, and many more fascinating properties. I hope you enjoy this illustrated guide and I encourage you to explore each of the formats more thoroughly on your own!

2" Quad

Developed by
Ampex

Era
1956–early 1980s

Format
analog

Capacity
1 hour

Size
12" reels

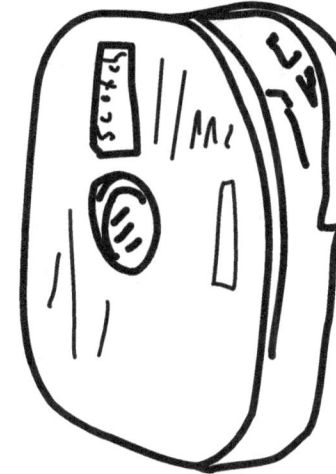

Fun fact

This format was the broadcast standard from its invention until the debut of 1" Type C in 1976

Fun fact

This format was used by broadcast television companies to air a show at the same time on East and West US coasts

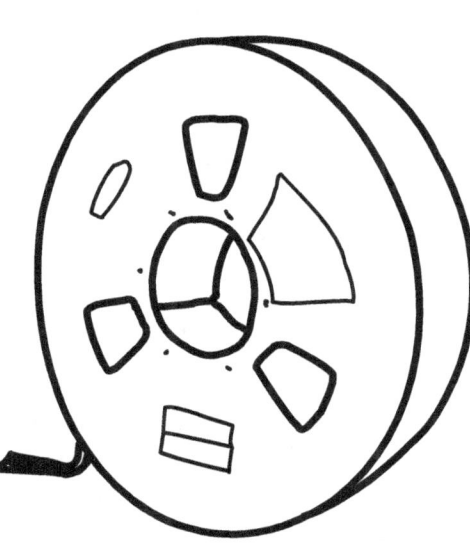

Fun fact

Most video formats used helical scanning, where video is scanned diagonally, but 2" Quad is based on quadrature scanning, a method where 4 video heads scan video tape at a 90° angle

2" Helical

Format
analog

AKA
**2 inch helical VTR,
VR-1500, VR-660**

Size
**6.5", 8", 10.5",
12.5" reels**

Developed by
Ampex

Era
1961–1970s

Capacity
5 hours

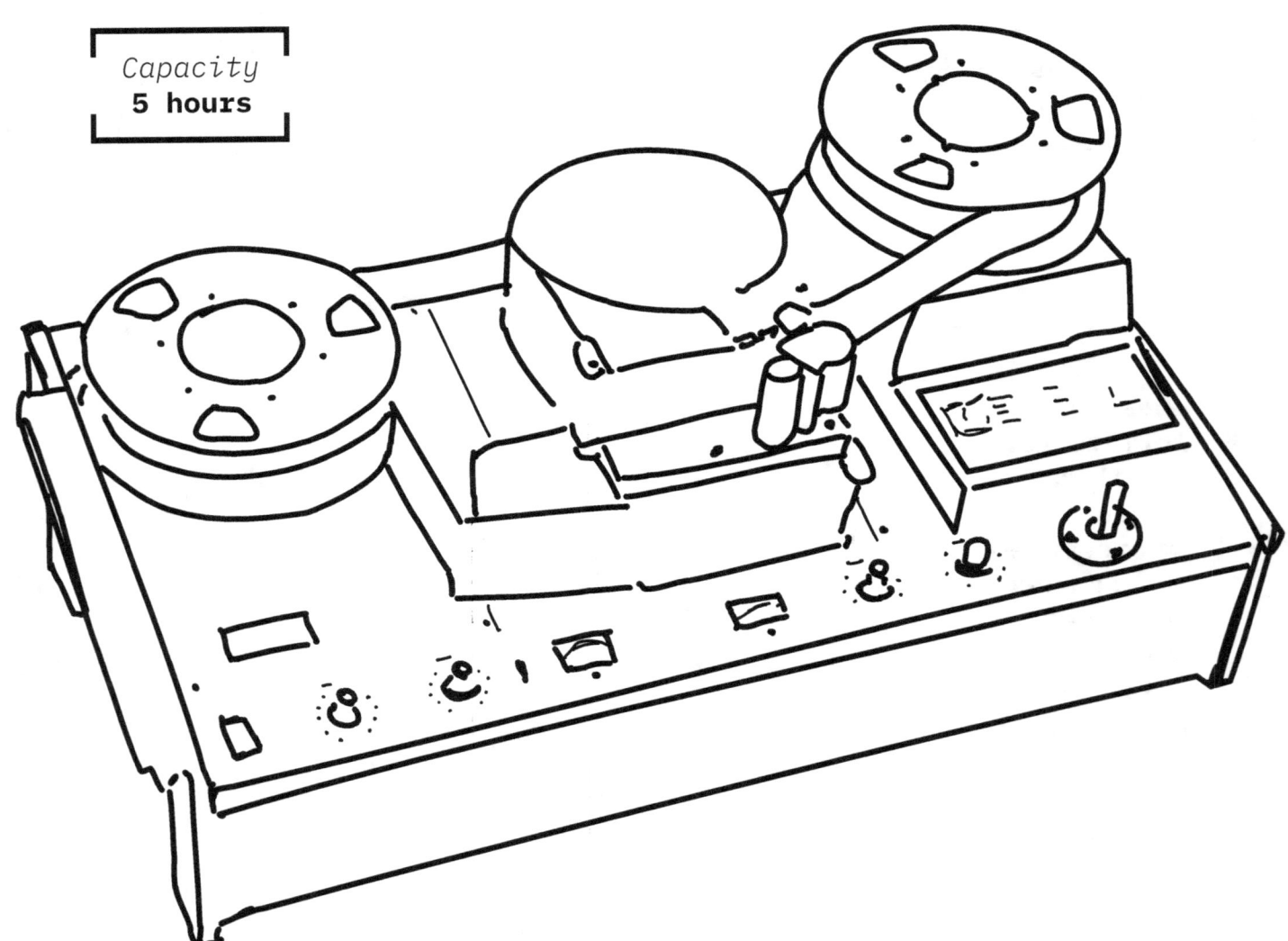

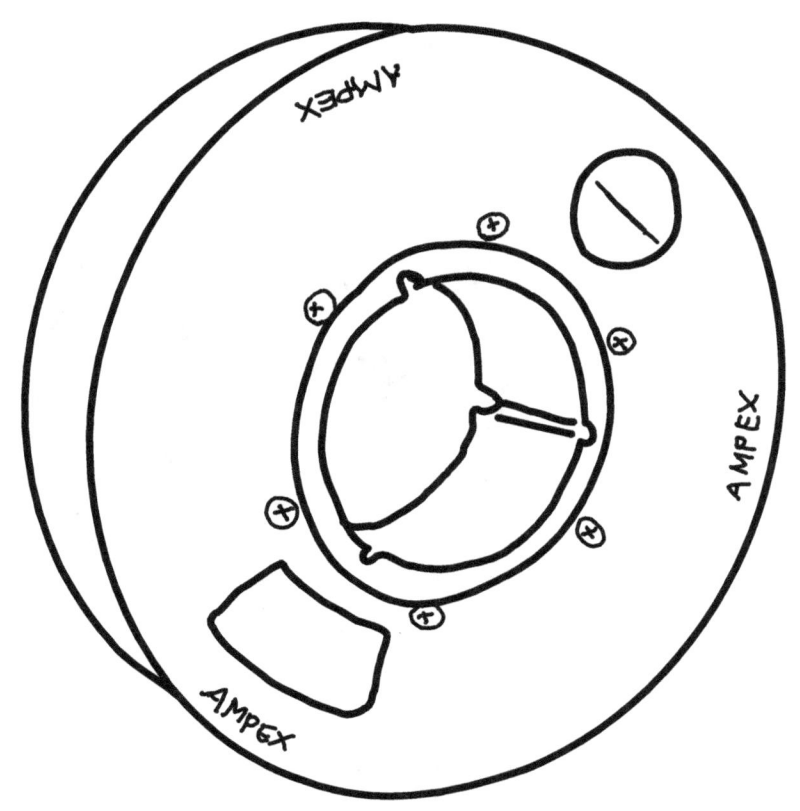

Fun fact
**The VR-660
(the professional
version of the 2"
helical scan deck)
weighed 100 pounds,
which was considered
light in comparison
with other video
recorders from the
same era**

Fun fact
**This format was not
suitable for broadcast
use and the machine
was too large and
pricy for consumers,
so this format was
marketed to industrial
and educational sectors**

Fun fact
**The 2" tape runs at 3.7
inches per second**

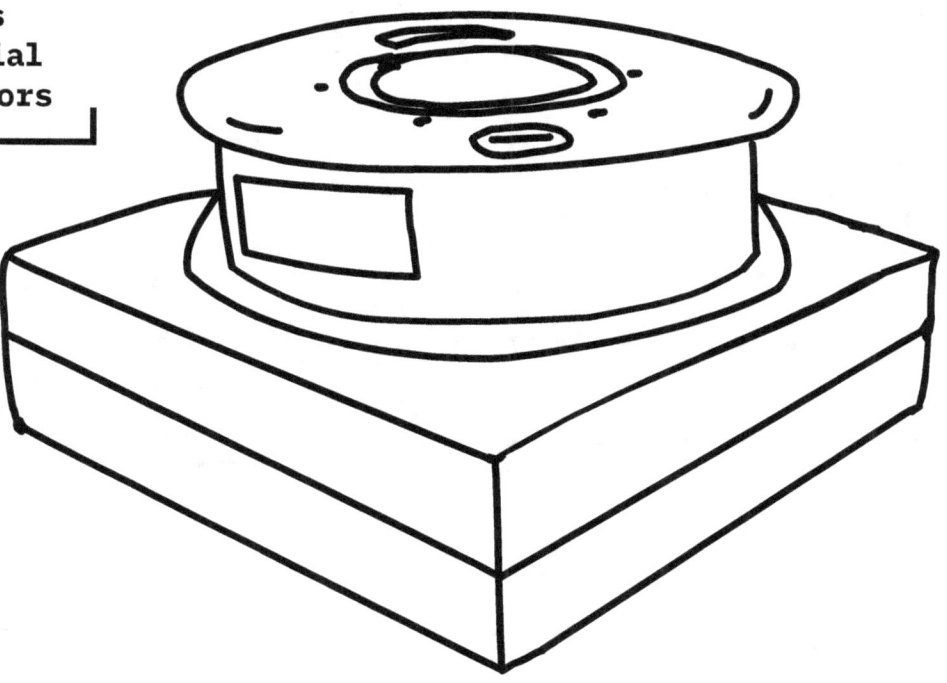

1/2" CV

Format
analog

AKA
1/2" open reel,
skip field

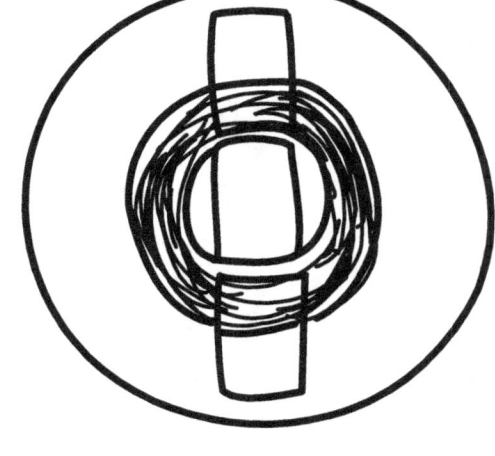

Era
1965–1970s

Developed by
Sony

Size
5" & 7" reels

Capacity
Small: 30 minutes
Large: 1 hour

Fun fact
This format introduced
the first portable
recording device,
the Porta-Pak

Fun fact
This format was much cheaper and smaller than previous open reel formats, and was the first format to target the non-professional and home use markets

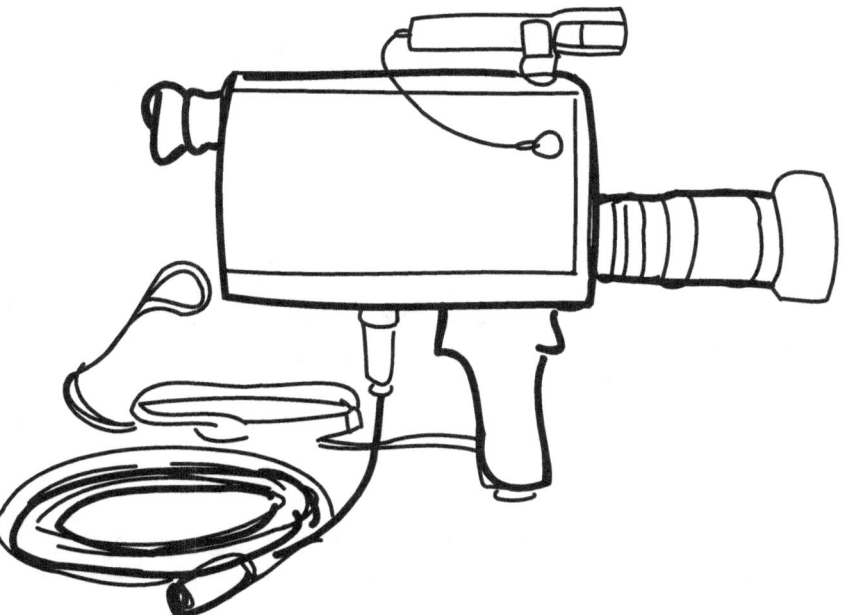

Fun fact
This format was also known as "skip field" because it recorded one field of video and repeated it twice during playback

1" Type A

Format
analog

Capacity
60 minutes

Era
1965–1970s

Fun fact
**1" Type A,
Type B, and
Type C are not
compatible
with each other**

Size
Up to 12" reels

Developed by
Ampex

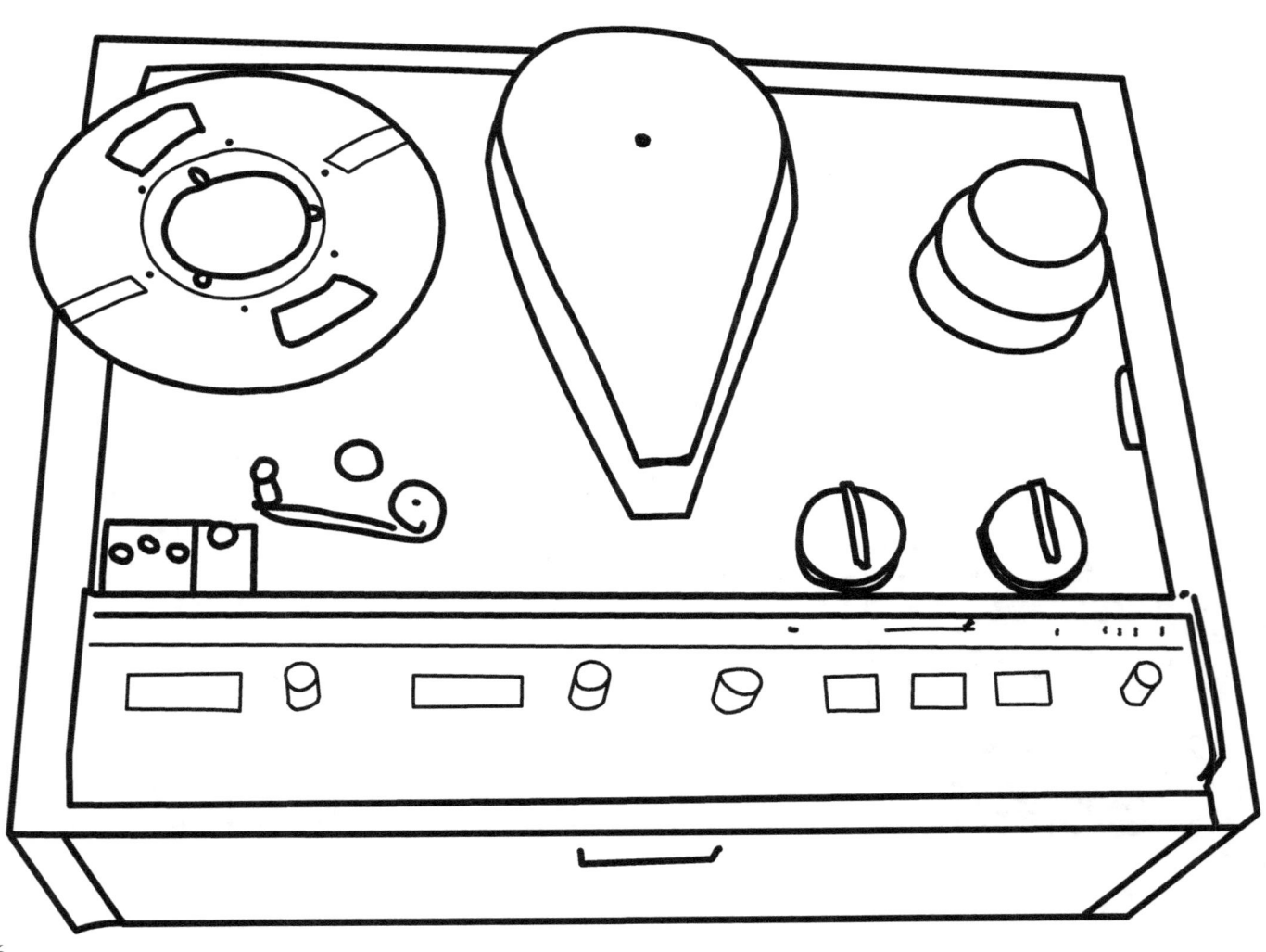

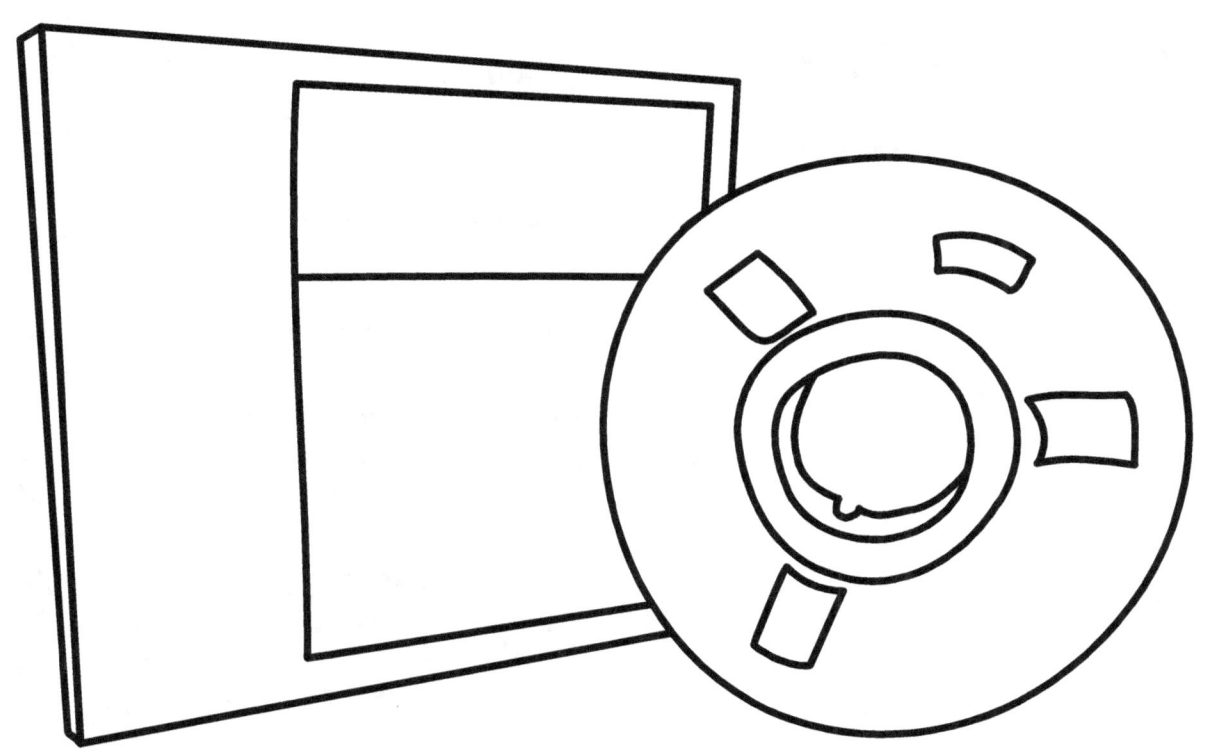

Fun fact
This format was
used by the
White House
Communications
Agency from
1966 to 1969

Fun fact
This format was not used
for broadcast television
because it did not meet
Federal Communications
Commission (FCC)
specifications for
broadcast videotape
formats

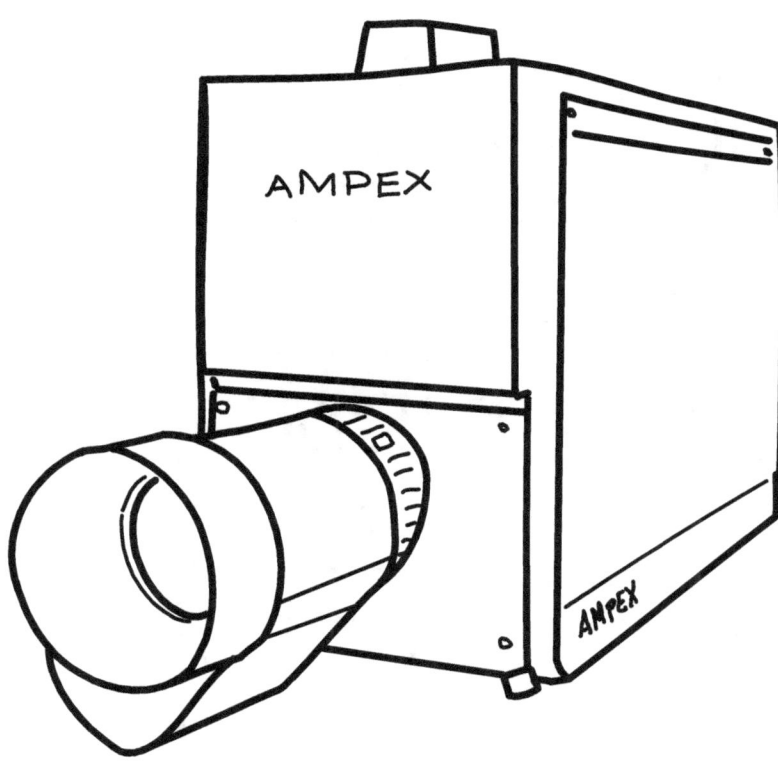

1" IVC

AKA
IVC 700, IVC 800, IVC 900

Era
1967–late 1980s

Capacity
1 hour

Size
8" reels

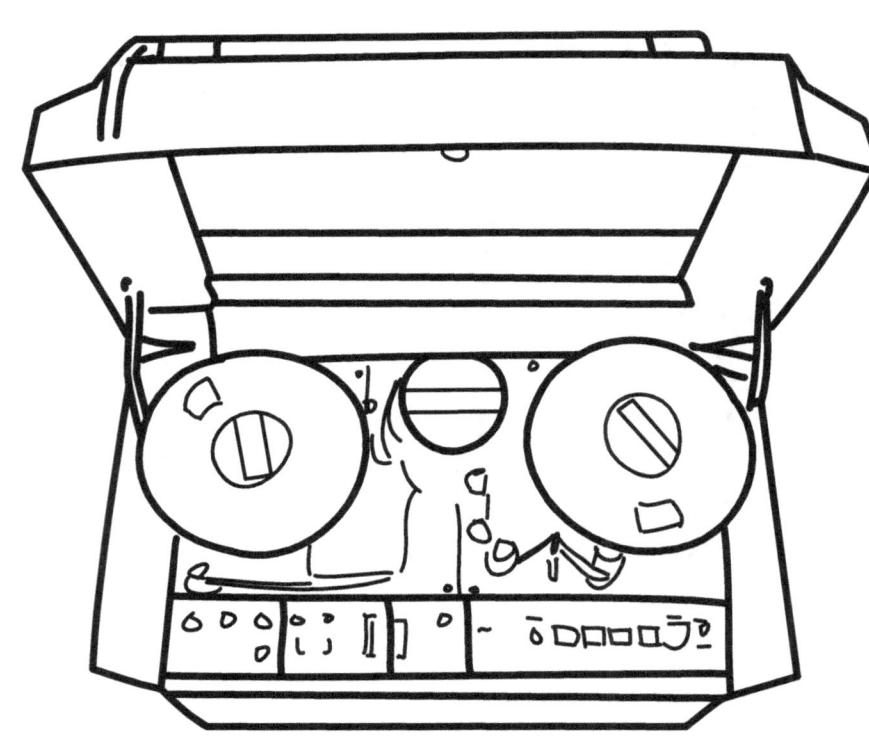

Developed by
International Video Corporation

Fun fact
These 1" tapes were produced by the International Video Corporation and came in a few different formats: 700, 800, 900, with the 800 series being the most popular; IVC also produced a 2" tape system, the IVC model 9000 VTR

Fun fact
IVC (UK Ltd.)
won a Royal
Television Society
award in 1984
for outstanding
new product of the
year and a Queen's
Award for
Technology
in 1985

Fun fact
This format
was largely
marketed in
the United
States and
United
Kingdom

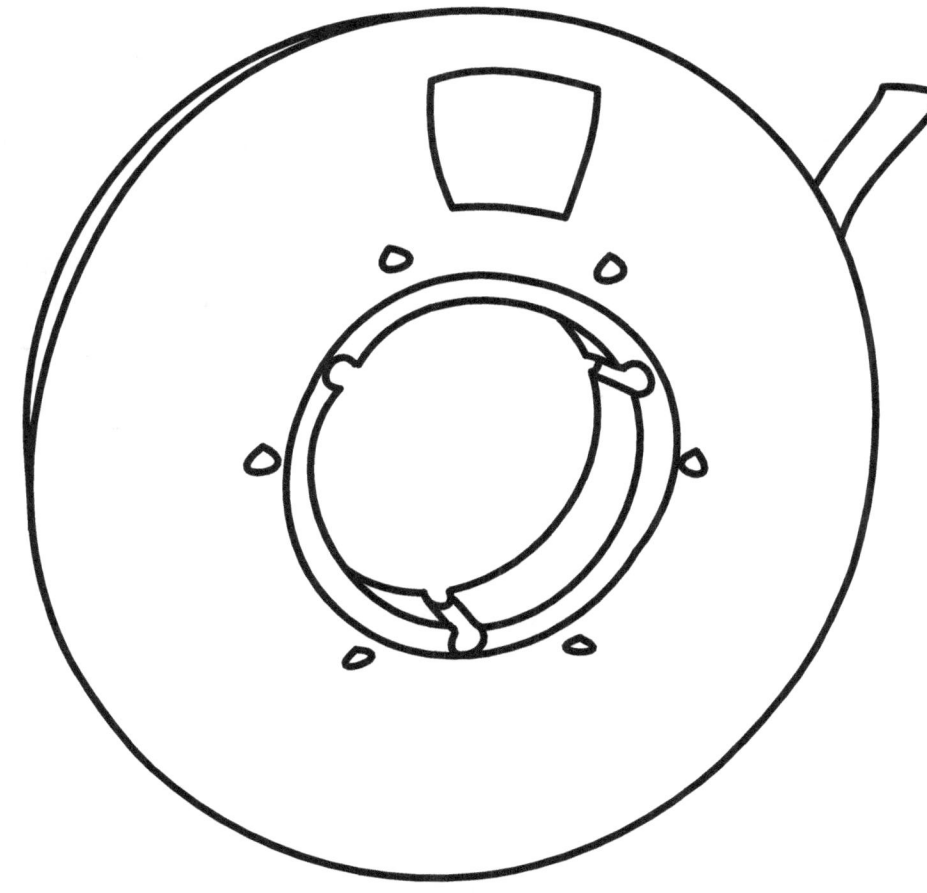

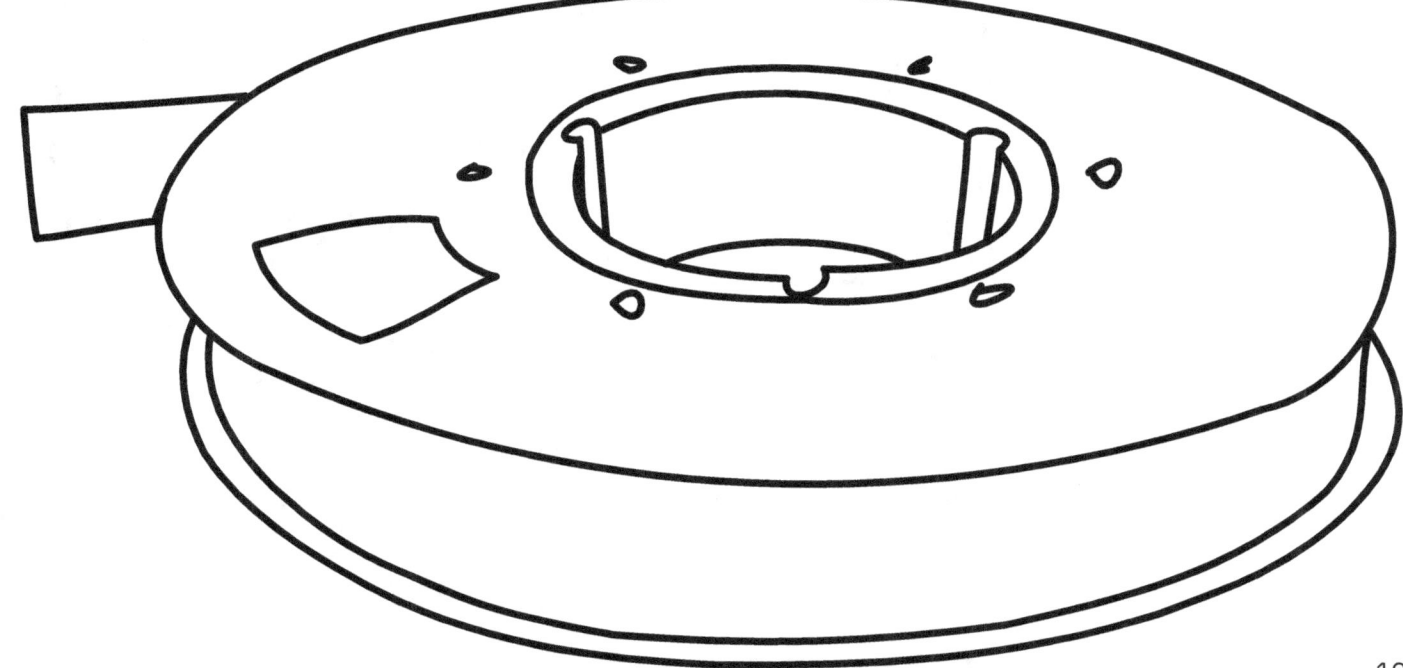

1/4" Akai

Developed by
Akai

Era
1967–late 1970s

Fun fact
This format was initially only available in black-and-white, with color being introduced in 1974; color tapes were not backwards-compatible due to the systems running at different speeds

Capacity
30 minutes

Size
5" reels

Format
analog

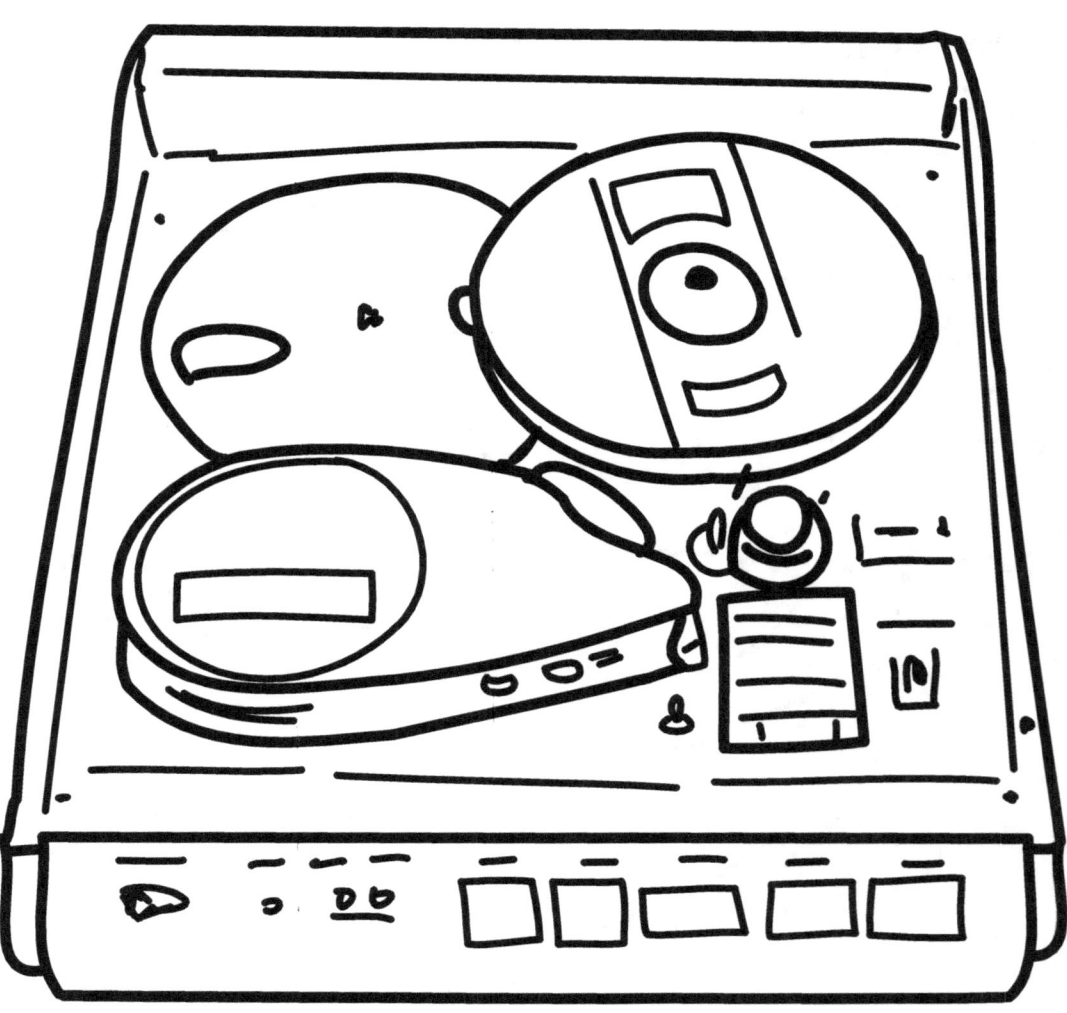

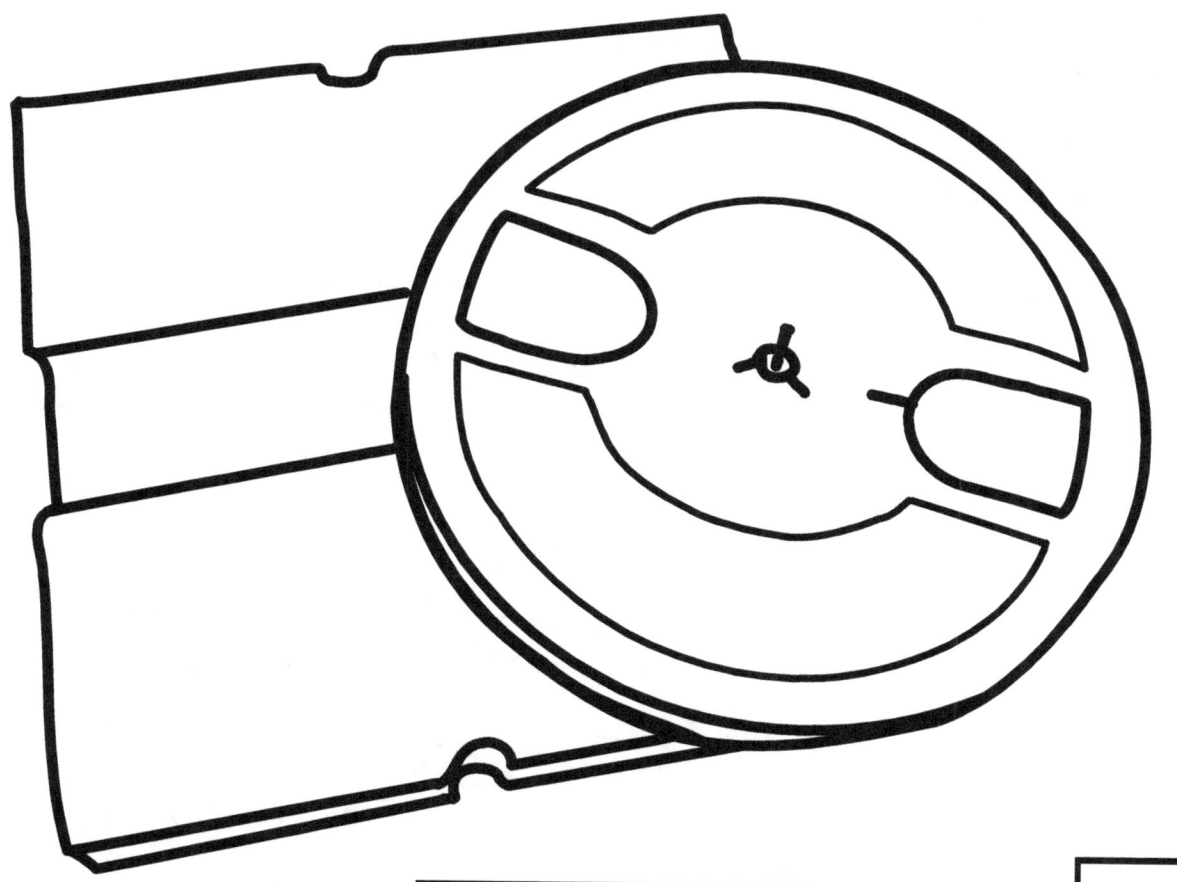

Fun fact
This format was typically stored on 5" reels but larger 10.5" reels were later introduced

Fun fact
This format was marketed for consumer use, intended for recording home movies

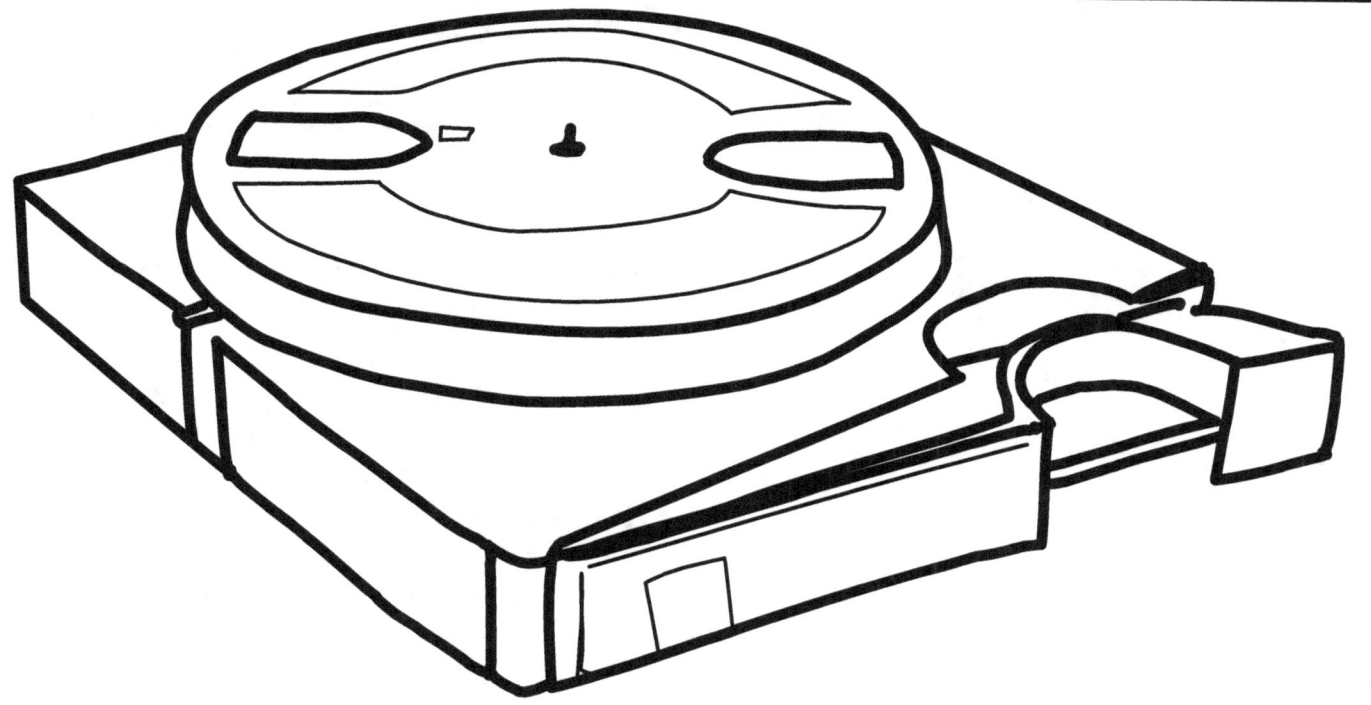

EIAJ

Format
analog

AKA
1/2" open reel

Era
1969-early 1980s

Developed by
Electronic Industries Association of Japan

Capacity
Small: 30 minutes
Large: 1 hour

Size
5" & 7" reels

Fun fact
This format was initially only in black-and-white, but color versions were developed later on

Fun fact
Despite originating with the 1/2" CV, this format became associated with the portable recording device (known as the PortaPak), which gave birth to the age of electronic news gathering

Fun fact
This format's success can be attributed to it having been developed as a shared, non-proprietary standardized format that could be developed and manufactured by different companies

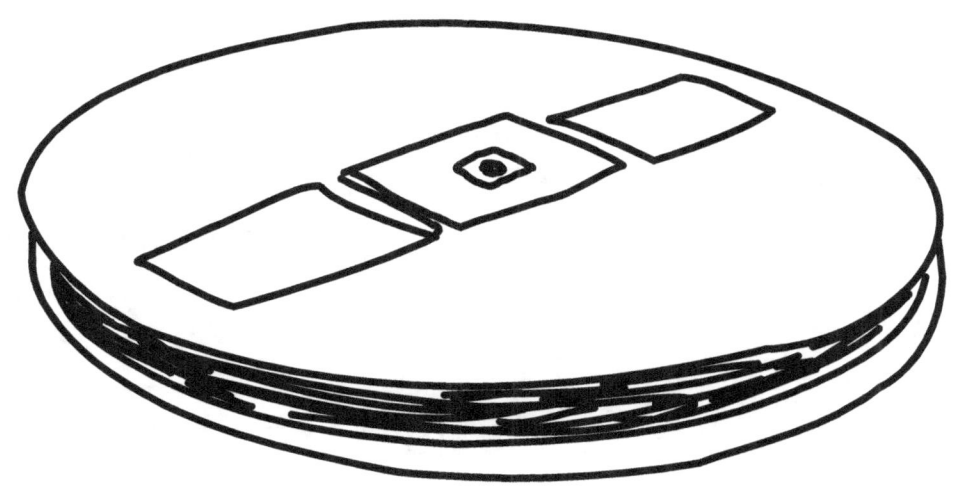

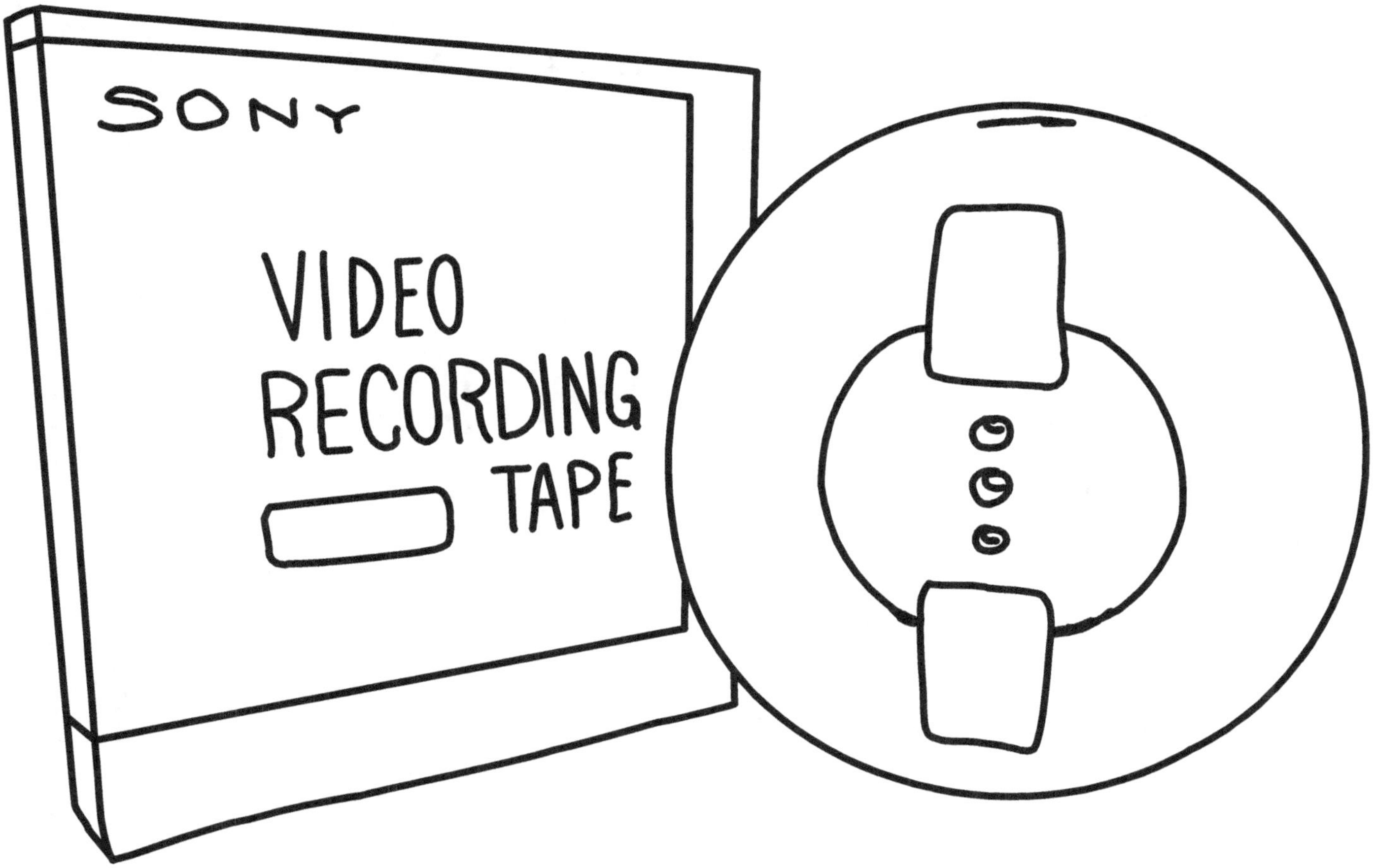

SONY

VIDEO RECORDING TAPE

3/4" U-matic

Format
analog

Era
1971–1990s

Developed by
Sony

Capacity
Small: 20 minutes
Standard: 60 minutes

Size
Small (S): 18.4 × 12.2 × 3.2 cm
Standard (SP): 22.1 × 14.0 × 3.2 cm

"S"

"SP"

"S SP"

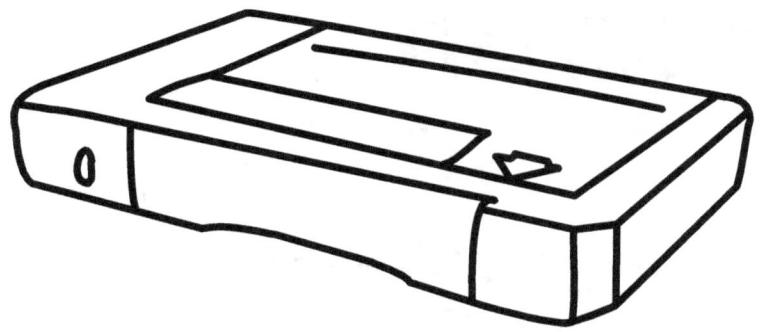

Cartrivision

Format
analog

Era
1972–1973

Developed by
Avco

Size
18 × 17 × 3.8 cm

Capacity
114 minutes

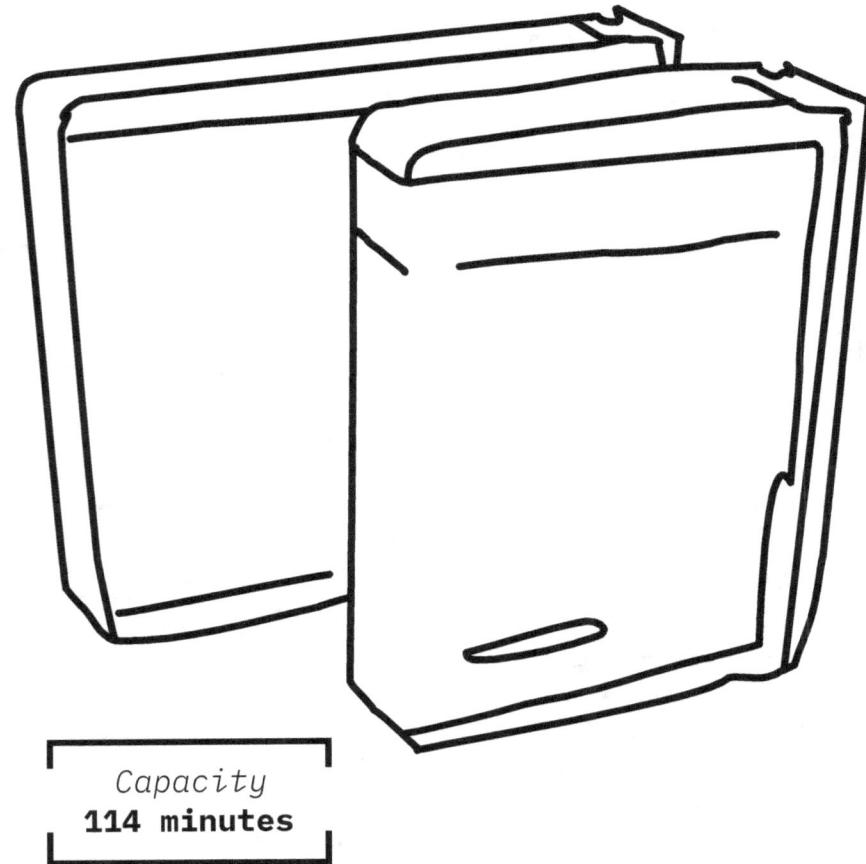

Fun fact

This format was on the market for only one year, from June 1972 to July 1973, before manufacturing ceased due to low sales

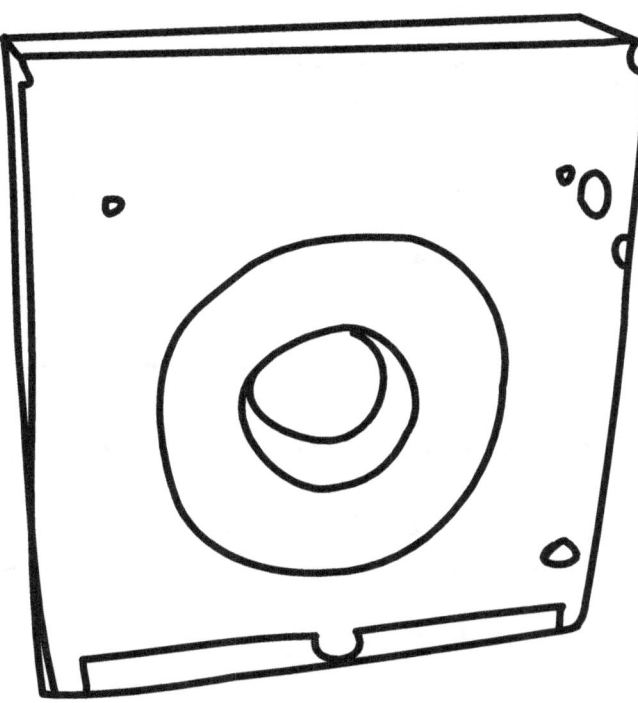

Fun fact

The first model of a Cartrivision -equipped TV set sold for (US) $1,350

Fun fact

Red cassettes, meant for video rentals, could not be rewound by a home machine

Video Cassette Recording

Era
1972–1977

Developed by
Philips

Format
analog

Size
12.7 × 14.5 × 3.8 cm

AKA
VCR, VCR-LP,
Super Video (SVR),
N1500/N1700

Capacity
VCR: 60 minutes
VCR-LP: 145 minutes
SVC: 240 minutes

Fun fact
This was the first
successful
consumer-level home
videocassette recorder
system

Fun fact
This format was marketed
only in the U.K., Europe,
Australia & South Africa

Fun fact
This format introduced many features adopted by later videotape format decks such as its control buttons style, clock with timer, and a built-in tuner

V-Cord/V-Cord II

Developed by
Sanyo

Era
1974–1977

Capacity
V-Cord: 1 hour
V-Cord II: 2 hours

Format
analog

Fun fact
Although the V-Cord II was based on V-Cord, which debuted in 1974, this sequel was too late to market in competition with Betamax and VHS; V-Cord manufacturers Sanyo and Toshiba later switched to producing Betamax

Size
11 × 15.7 × 2.5 cm

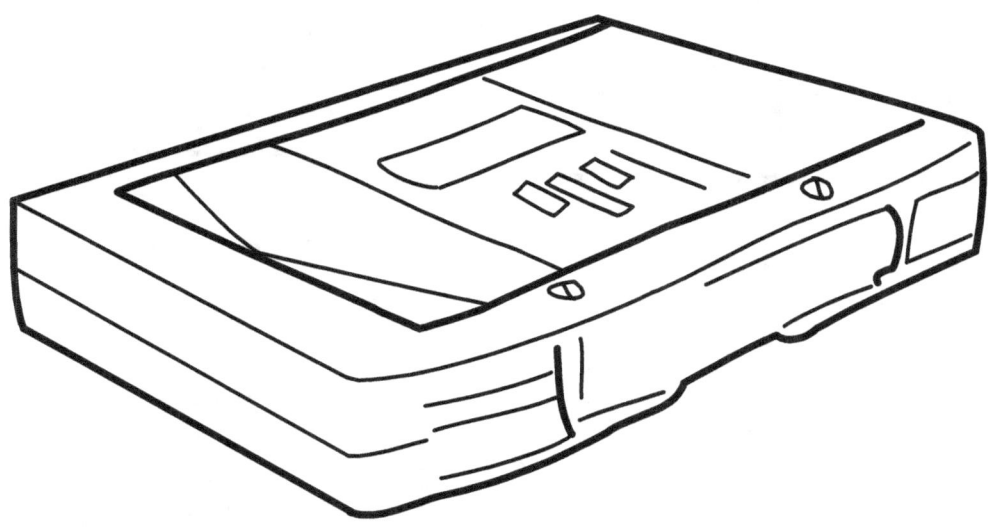

Fun fact
The original V-Cord cassette could record up to 60 minutes of black-and-white video and the V-Cord II could record up to 120 minutes in color

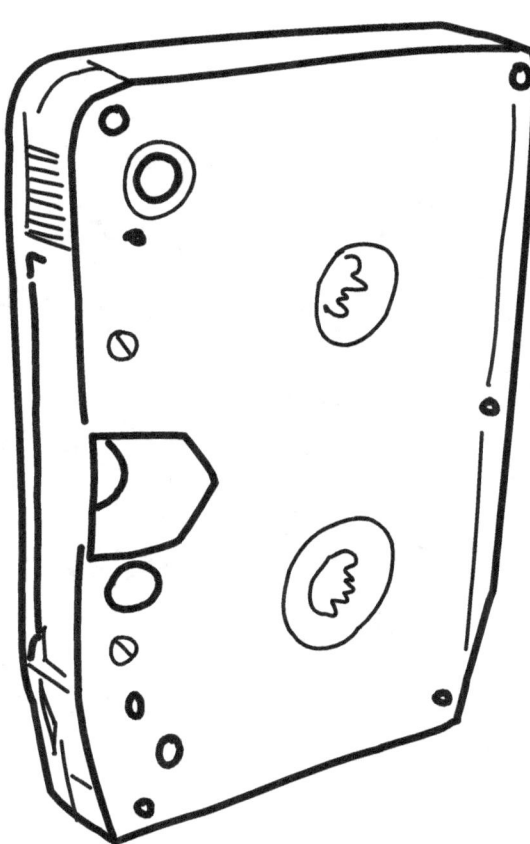

Fun fact
V-Cord II machines were the first consumer playback devices to offer two recording speeds: standard mode (STD), and a long-play mode (LP)

Betamax

Format
analog

Capacity
BI: 90 minutes
BII: 200 minutes
BIII: 5 hours

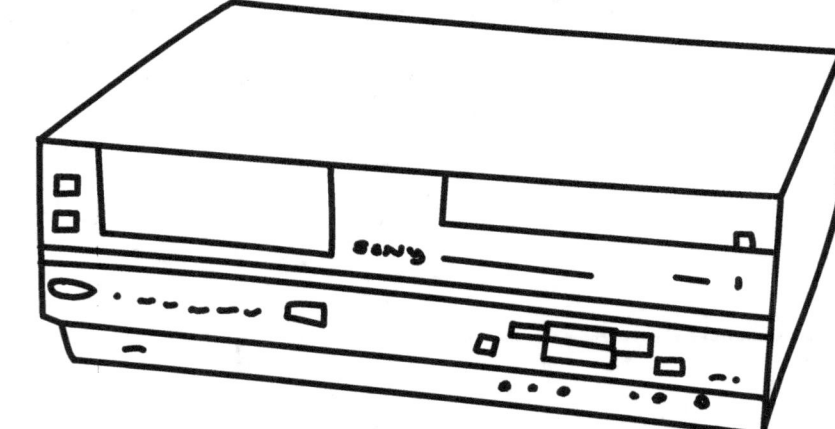

Size
15.6 × 9.6 × 2.5 cm

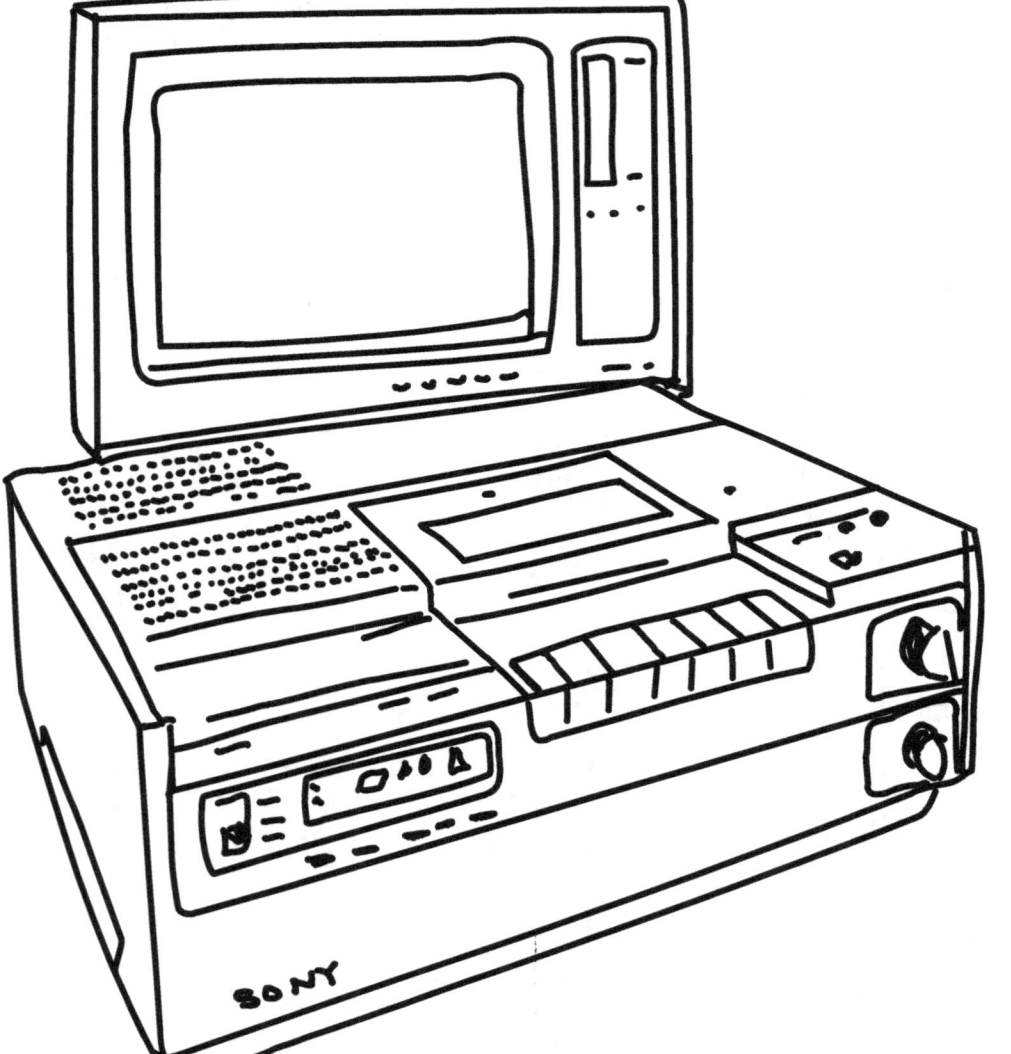

Era
1975–late 1980s

Developed by
Sony

Fun fact
This was the central format in a U.S. Supreme Court case that determined home videotaping of television and movies was legal, ruling against film industry corporations

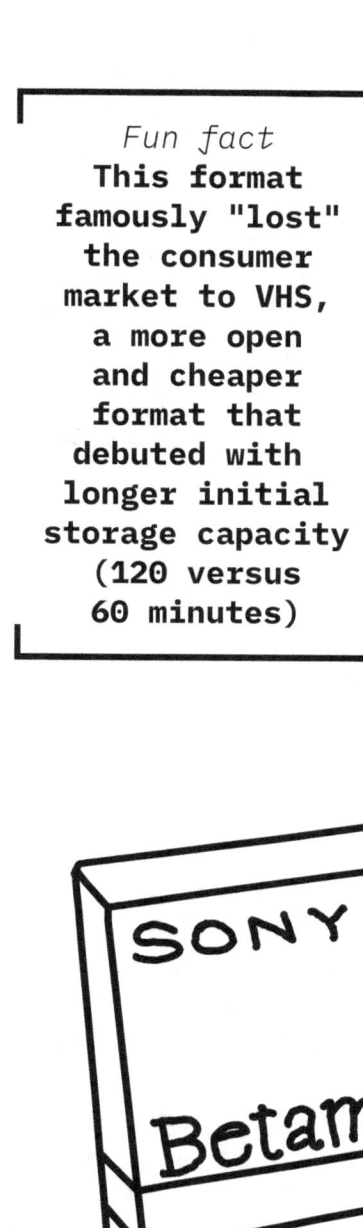

SONY

Betamax

1" Type B

Format	AKA	Era
analog	**B-Format**	**1975–1980s**

Developed by	Capacity	Size
Bosch	**2 hours**	**Up to 10.5" reels**

Fun fact
This format can be distinguished from the other 1" types by the way the tape wound, with the oxide facing out instead of in; Type C tape would face in and Type A could be either if created before 1975

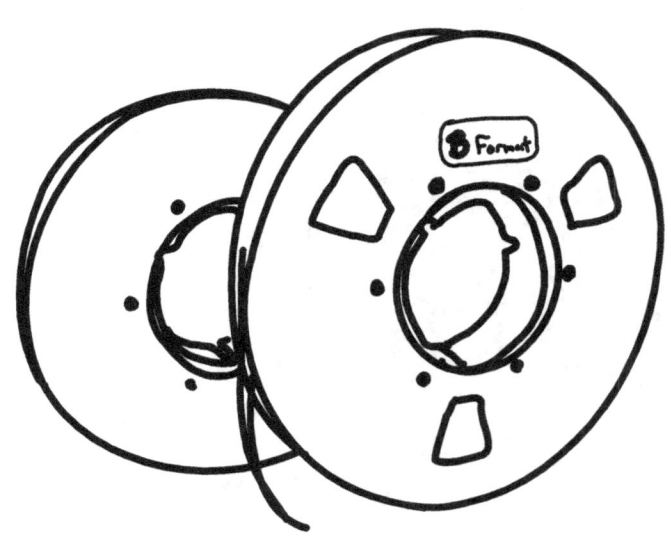

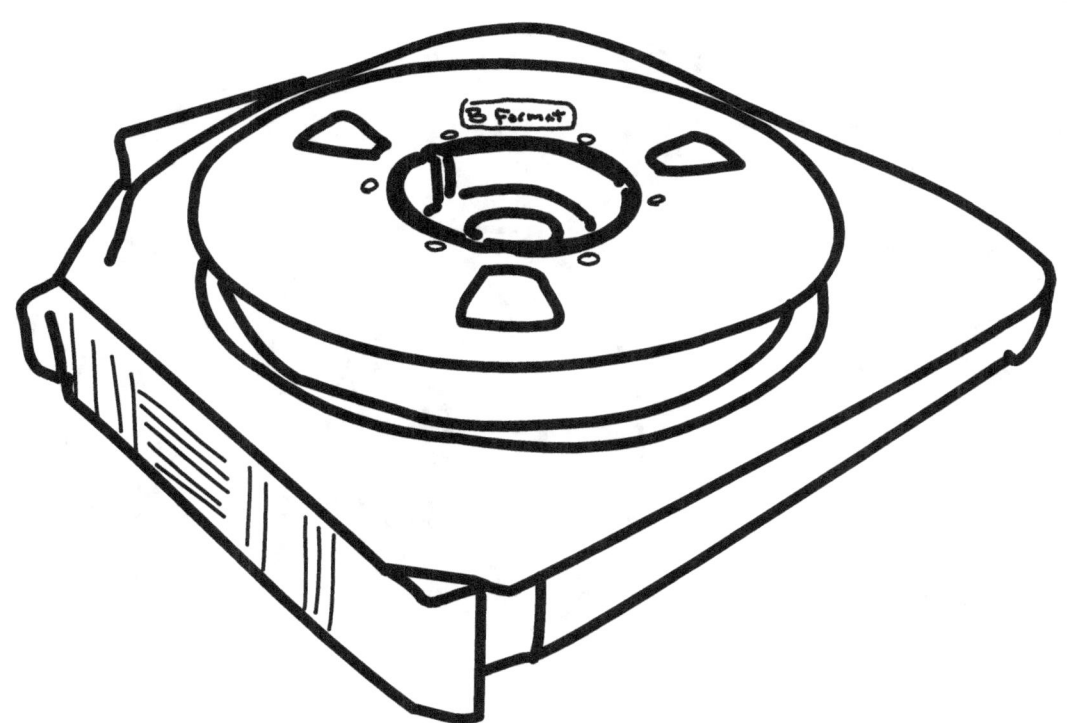

Fun fact
This format began to replace 2" Quad and 1" Type A, but it was later largely replaced by 1" Type C for use in broadcast television

Fun fact
This format segments each field to 5 (NTSC) or 6 (PAL) helical scan tracks per field, instead of recording one field per track

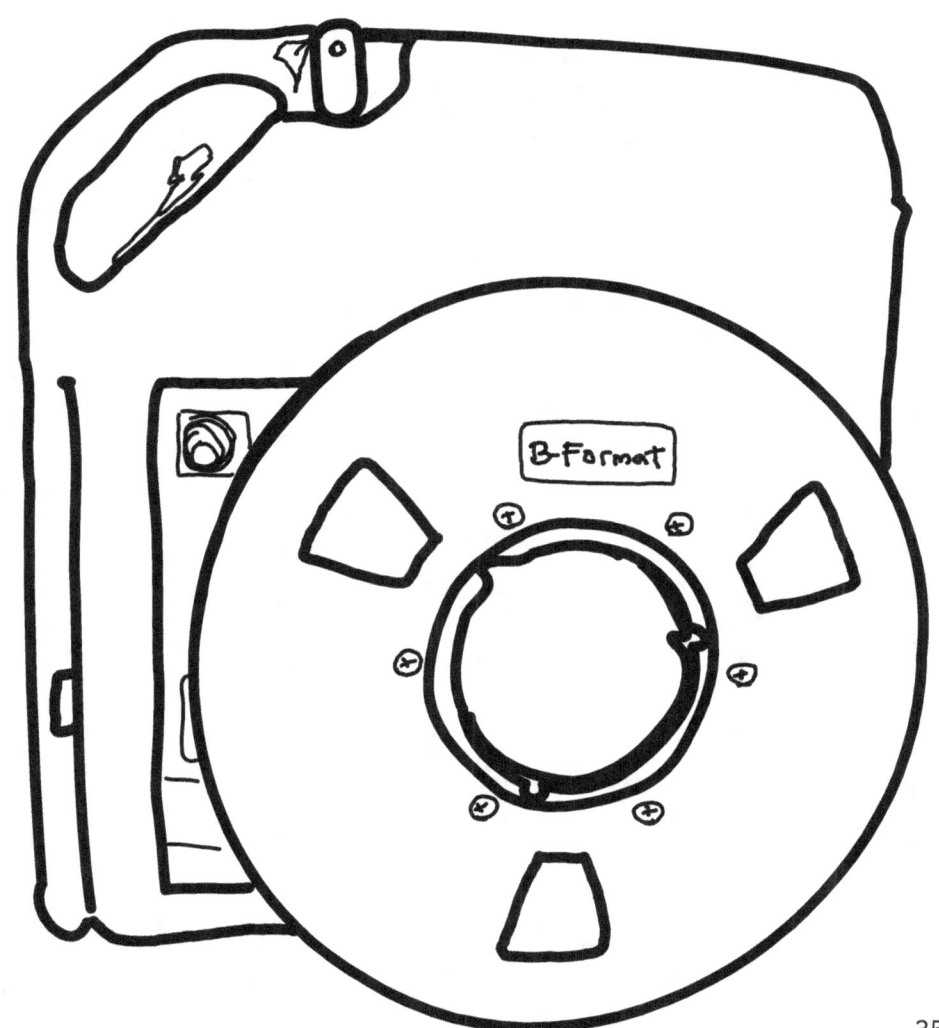

1" Type C

Era
1976–1990s

Developed by
Ampex & Sony

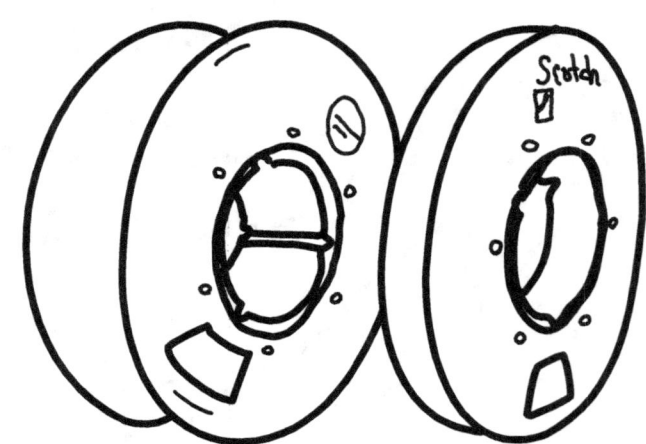

Capacity
Generation 1: 1 hour
Generation 2: 2 hours

Size
12" reels

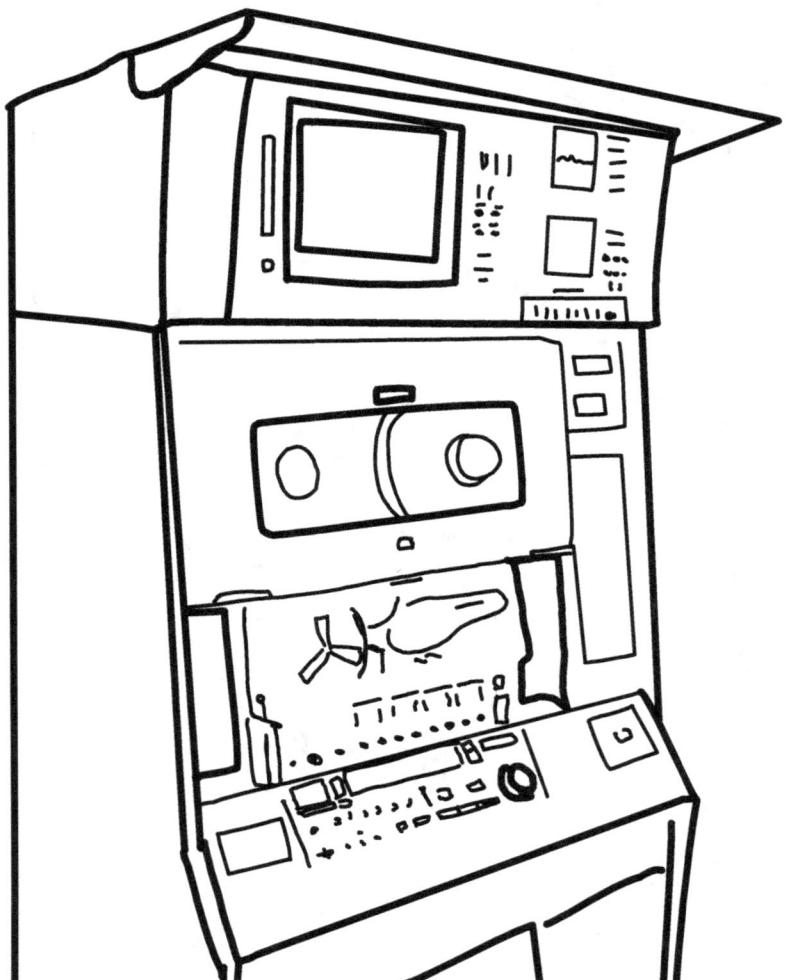

Fun fact
The most common size for this format was 12" reels, but smaller reels were produced for delivering shorter content

Fun fact
This format was capable of "trick-play" functions such as still frames, shuttle, and slow motion variable-speed playback

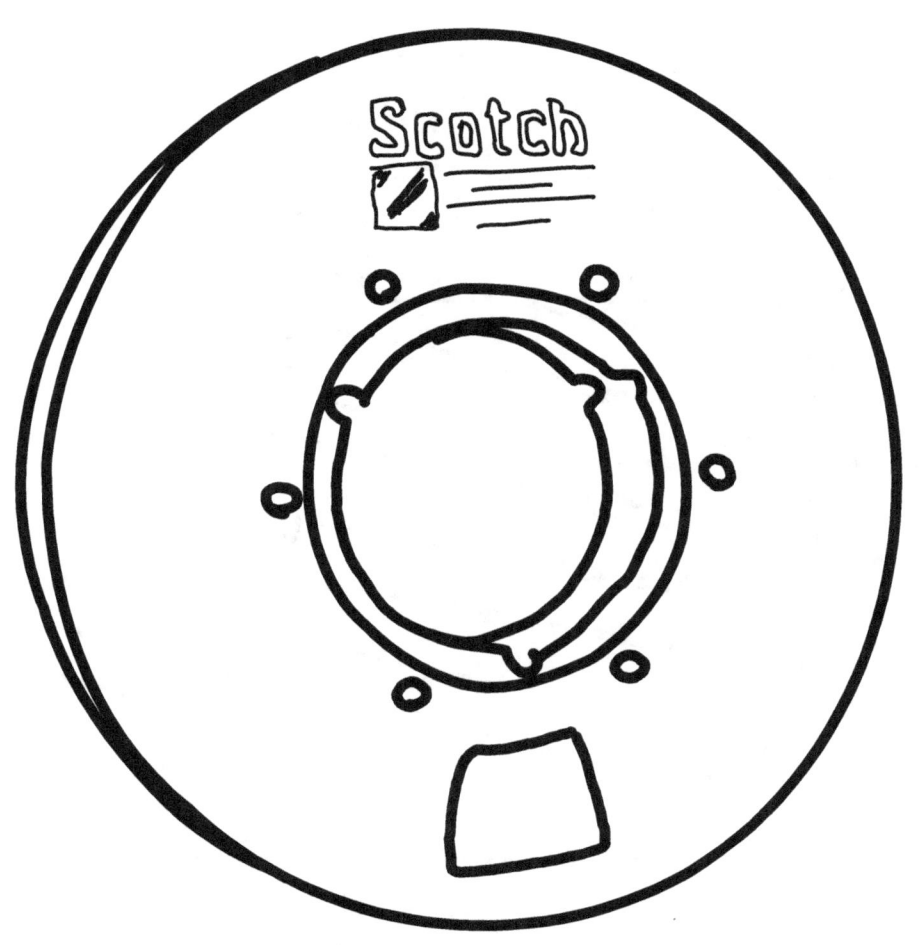

Fun fact
**Recording
devices
were able
to capture
single frames,
which made
this format
popular to
use during the
development
of early
computer
animation**

VHS

AKA
Video Home System

Format
analog

Developed by
JVC

Size
18.7 × 10.2 × 2.5 cm

Era
1976–2000s

Capacity
VHS-C: 30 minutes
Standard Play (SP): 2 hours
Long Play (LP): 4 hours
Extended Play (EP): 6 hours
Super Long Play (SLP): 6 hours

Fun fact
A smaller variant of VHS, VHS-C, was introduced in 1982 for use in camcorders; these tapes could be played in standard VHS machines with an adapter

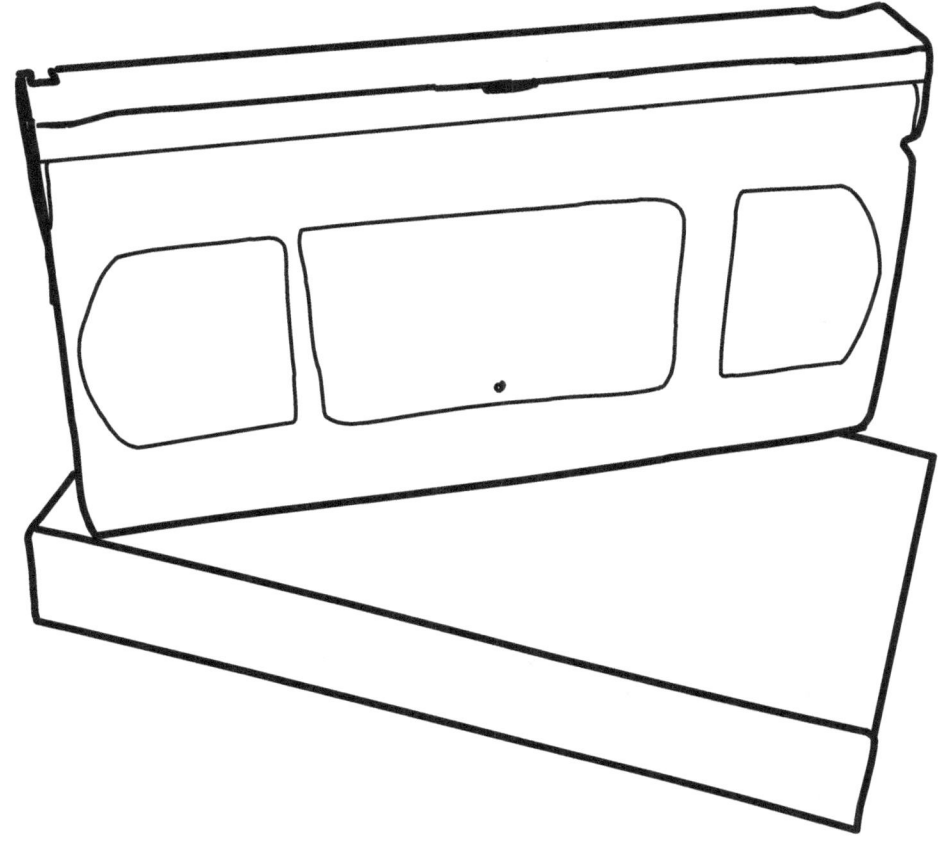

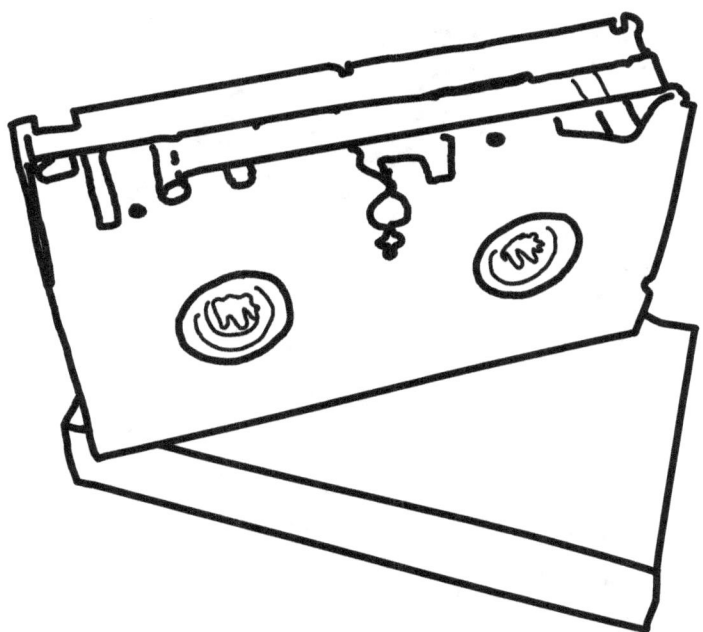

LaserDisc

AKA
LD, MCA DiscoVision

Format
analog

Era
1978-2001

Size
12" discs

Capacity
**1 hour
(per side)**

Developed by
**Philips, MCA Inc.,
Pioneer Corporation**

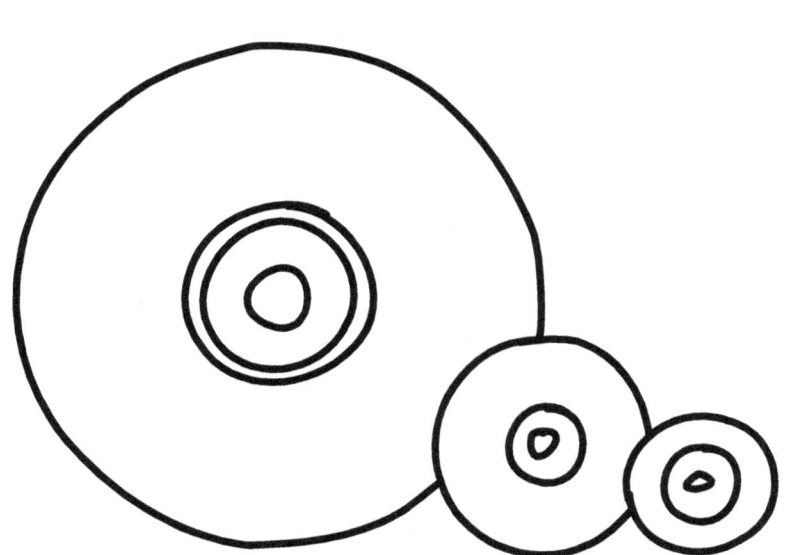

Video2000

AKA
Video Compact Cassette

Era
1979–1988

Developed by
Philips and Grundig

Fun fact
This format failed to gain commercial success due to competition with VHS & Betamax, the struggle to standardize players and tapes between companies, and higher cost

Size
10.8 × 7.2 × 2.1 cm

Capacity
8 hours (per side)

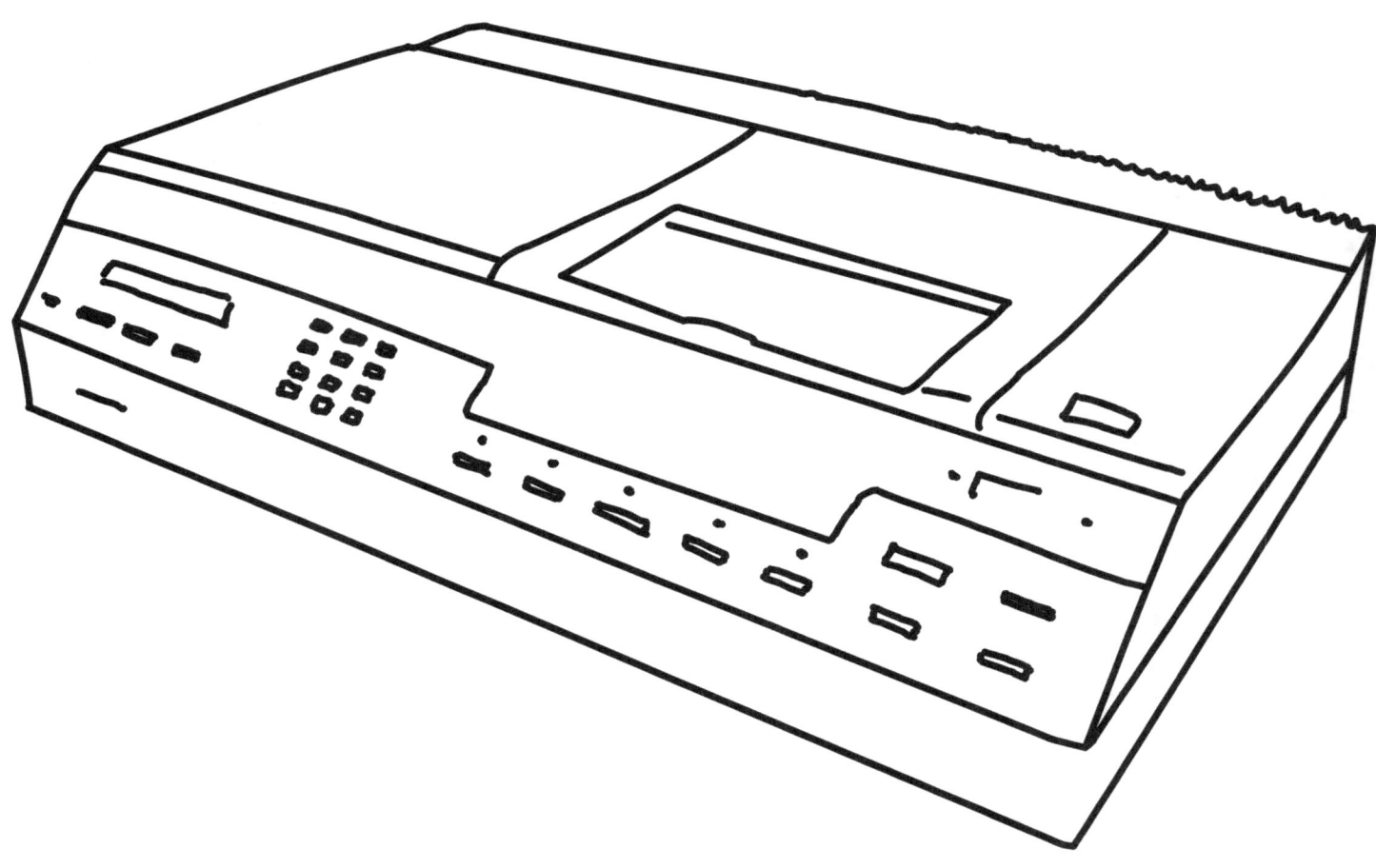

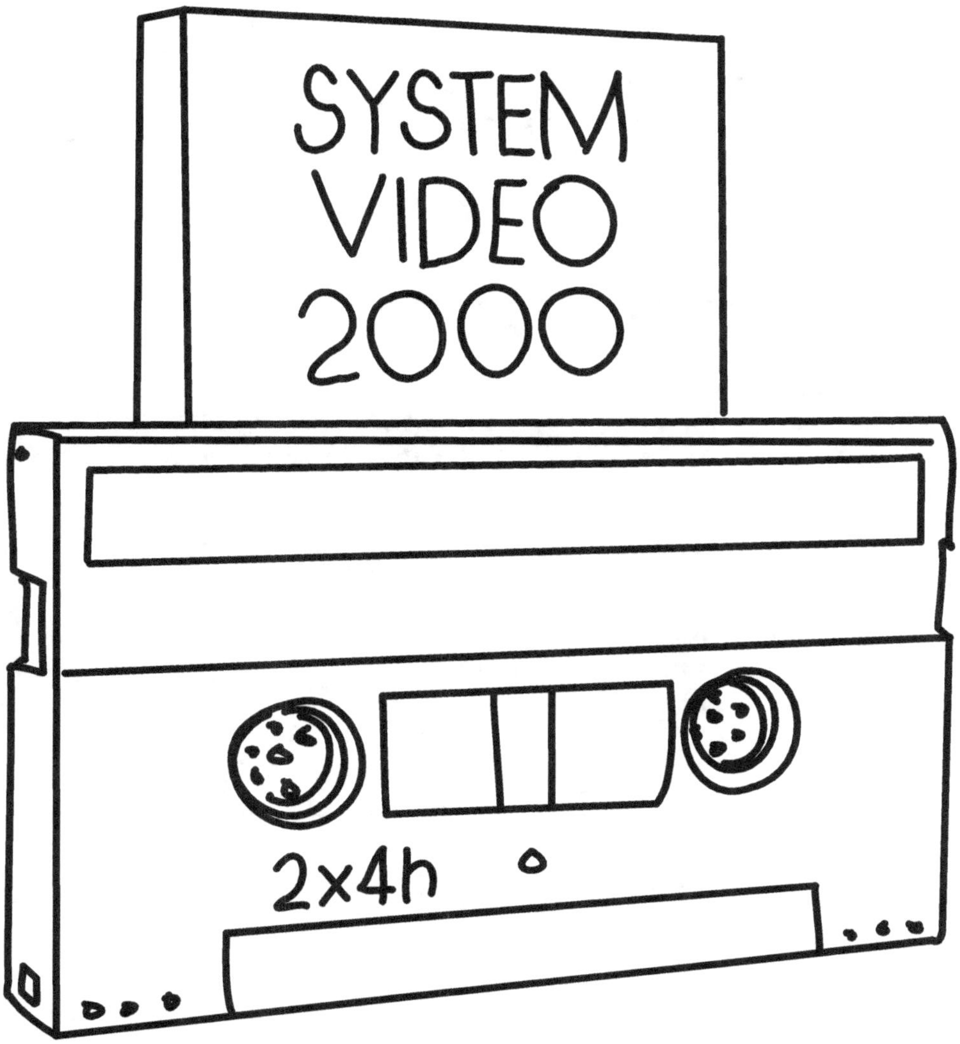

SYSTEM VIDEO 2000

2x4h

Fun fact
This format could play video in either direction and at any speed

Fun fact
This format could be turned over (like audio cassette tapes) to play the tape on both sides

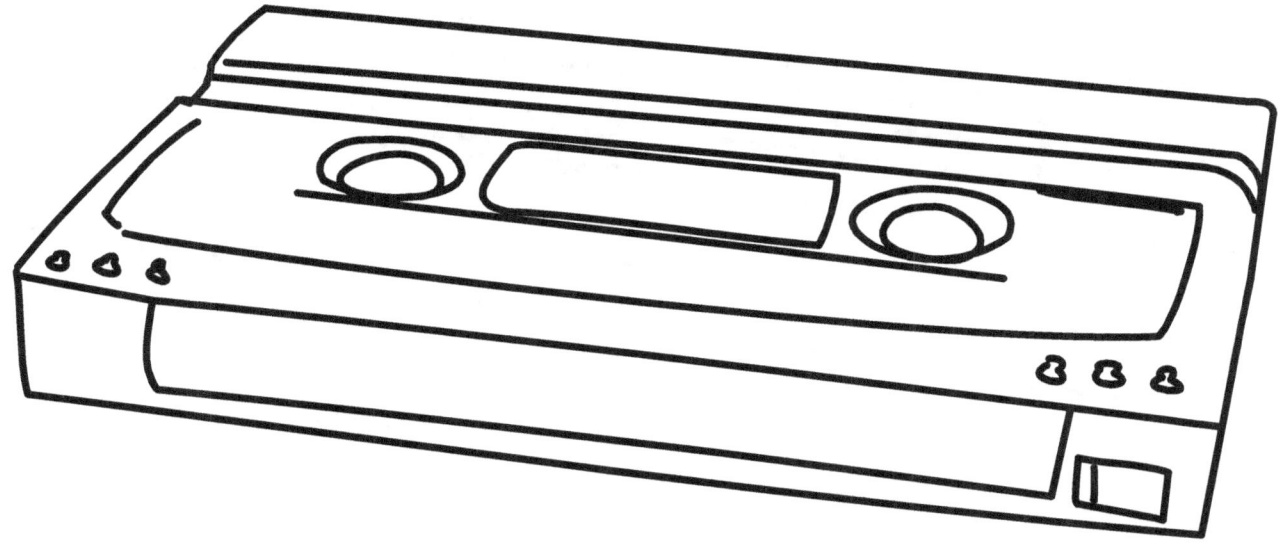

43

CED

AKA
Capacitance Electronic Disc, SelectaVision

Format
analog

Developed by
RCA

Capacity
1 hour (per side)

Fun fact
This format used a special stylus and high-density groove system similar to audio phonograph records; the discs were stored in a caddy from which the disc would be extracted by the playback device

Size
12" encased discs

Era
1981–1986

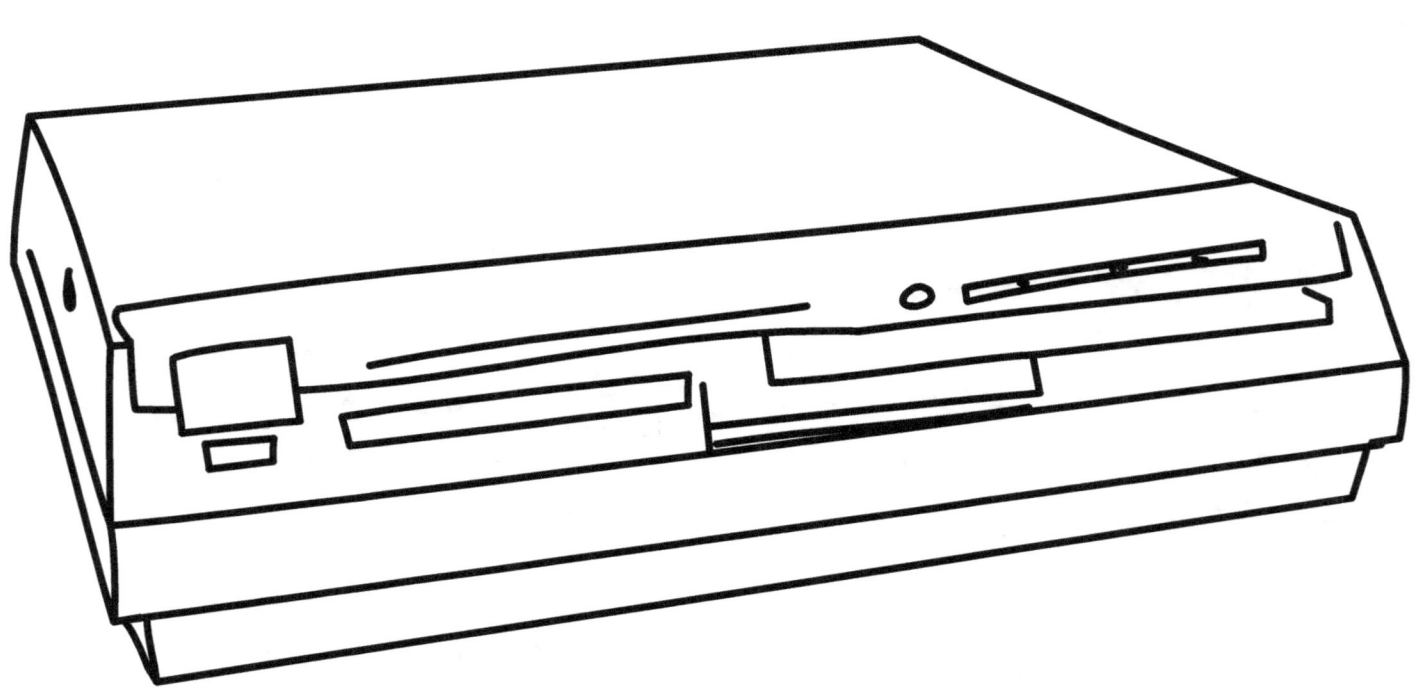

Betacam

Format
analog

Era
1982–2006

Developed by
Sony

Size
Short: 15.6 × 9.6 × 2.5 cm
Long: 25.4 × 14.5 × 2.5 cm

Capacity
Short: 63 minutes
Long: 194 minutes

Fun fact
This format is a higher-quality successor of Betamax intended for professional broadcast use instead of home consumer use; unlike Betamax, Betacam became the dominant broadcast format of its era & market

Fun fact
A blank Betamax tape can work on a Betacam deck and a Betacam tape can be used to record in a Betamax deck (although this practice is not recommended)

Video8/Hi8

Capacity
2 hours

Size
9.5 × 6.2 × 1.5 cm

Fun fact
This format's appeal was around its smaller size, which made for much lighter camcorders than VHS or Betamax

Era
Video8: 1985–2000s
Hi8: 1989–2007

Developed by
Sony

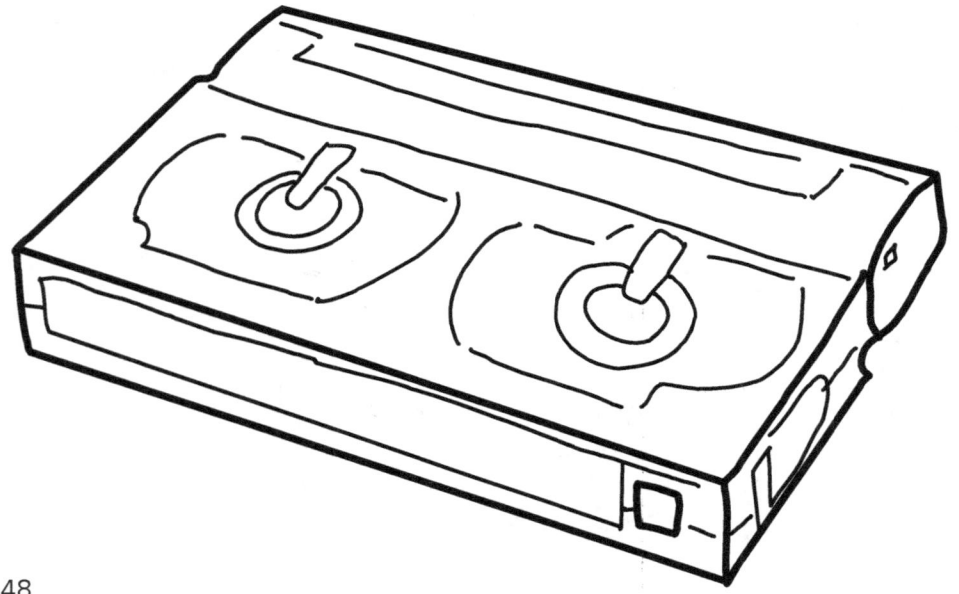

Fun fact
This format was initially launched by Kodak but Sony ended up dominating the market with its Handycam camcorder line

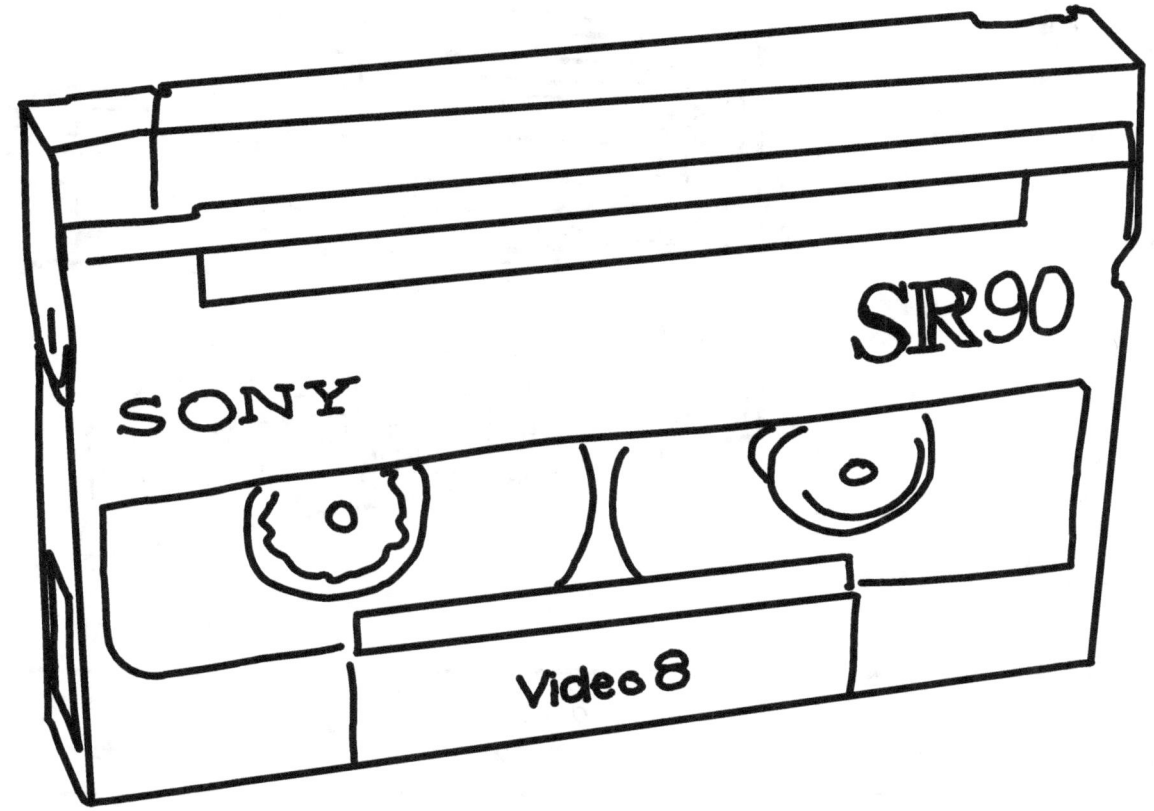

D-1

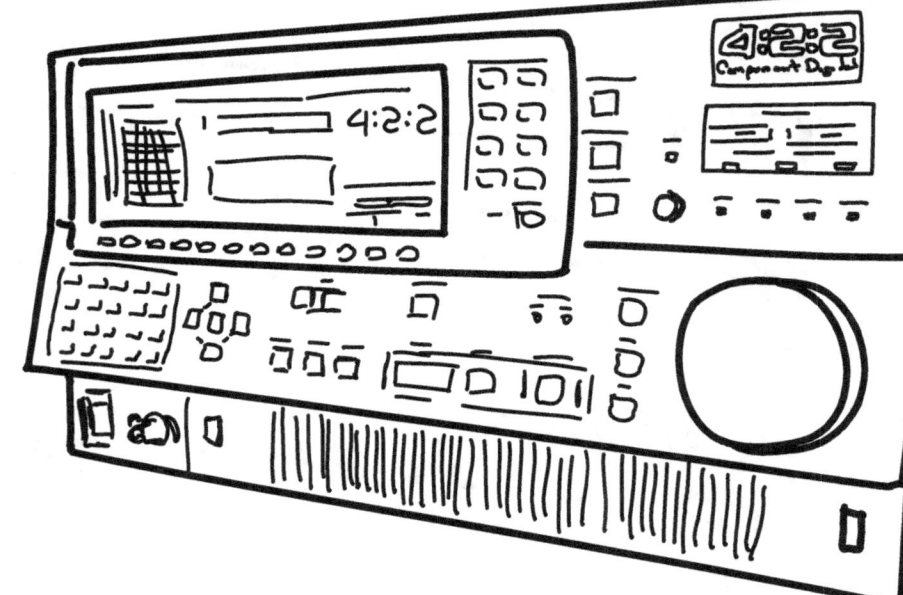

Format
digital

Developed by
Sony

Size
36.5 × 20.3 × 3.2 cm

Capacity
94 minutes

Era
1986–1990s

AKA
4:2:2 Component Digital

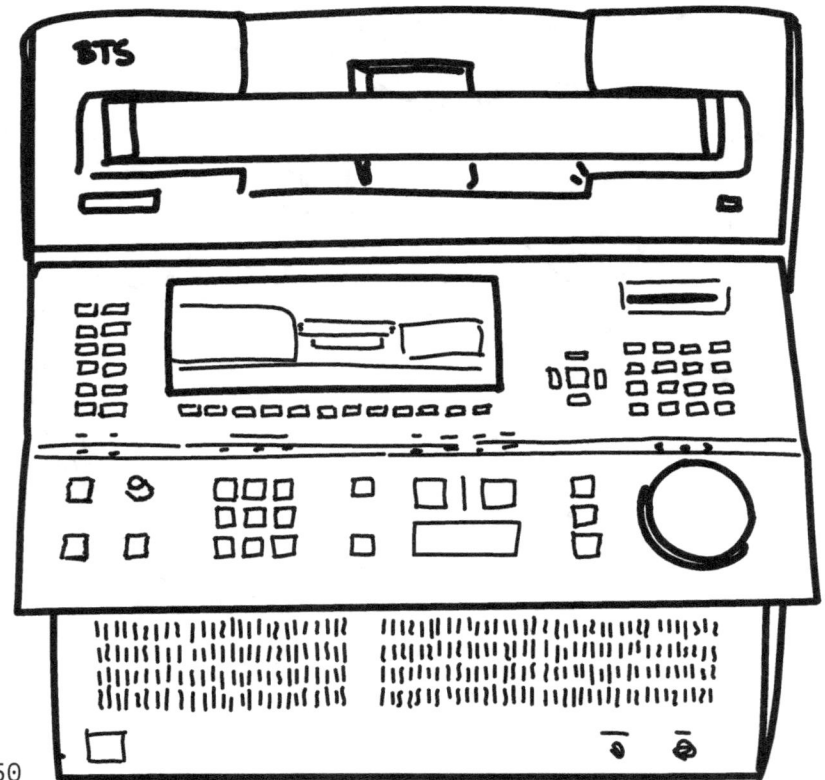

Fun fact
This format used component video (where video signals move through 3 separate cables), as opposed to the more common composite video (where the video signals are sent as one signal)

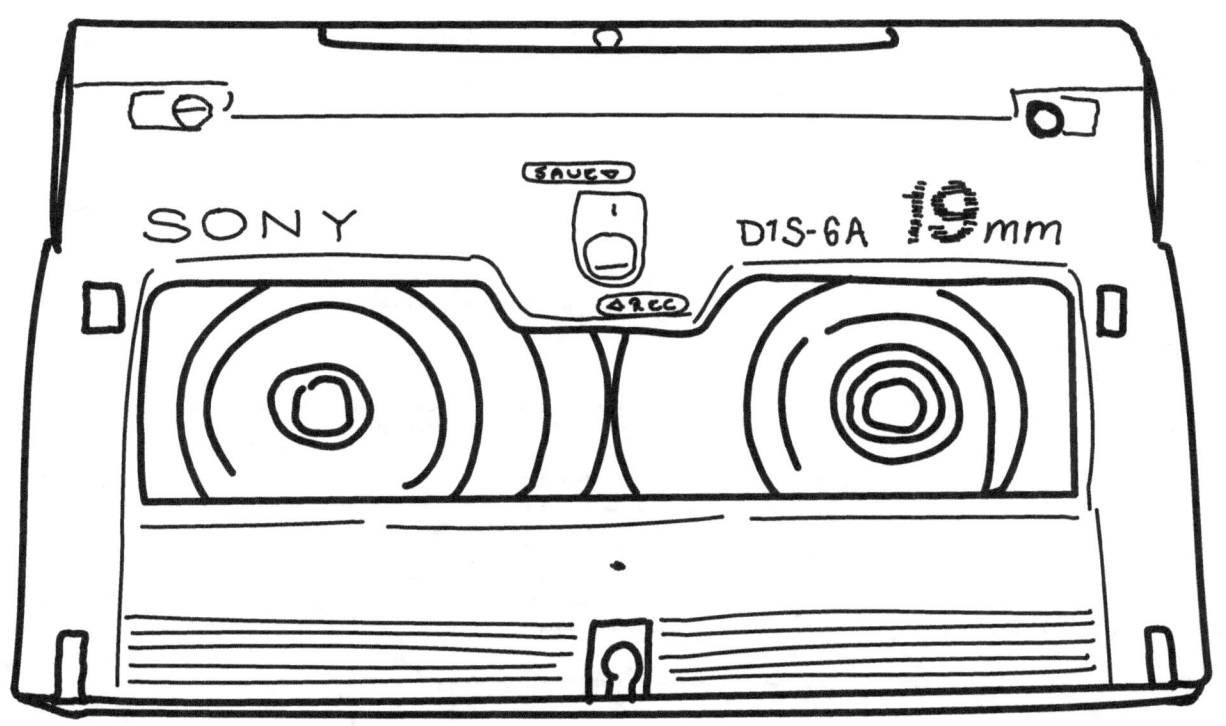

Fun fact
This format was the first major professional digital video tape recording system

Fun fact
This was the format of choice for animation studios

MII

Format
analog

Developed by
Panasonic

Era
1986–1990s

Capacity
Small: 20 minutes
Standard: 90 minutes

Size
Small: 13 × 8.7 × 2.5 cm
Standard: 18.7 × 10.6 × 2.5 cm

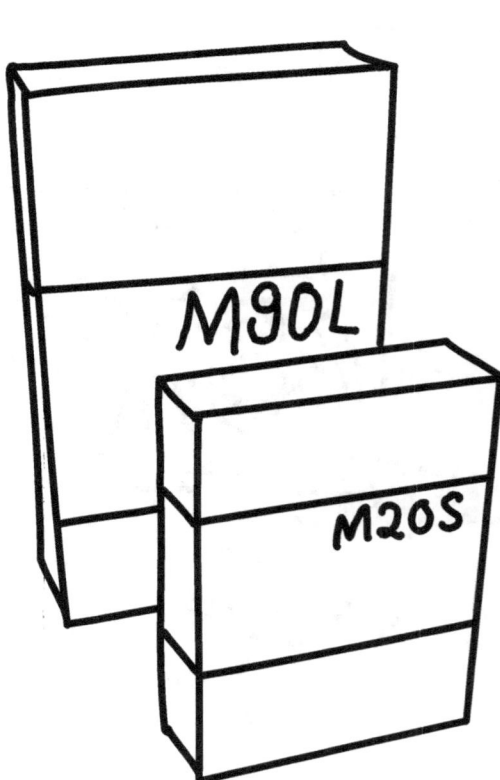

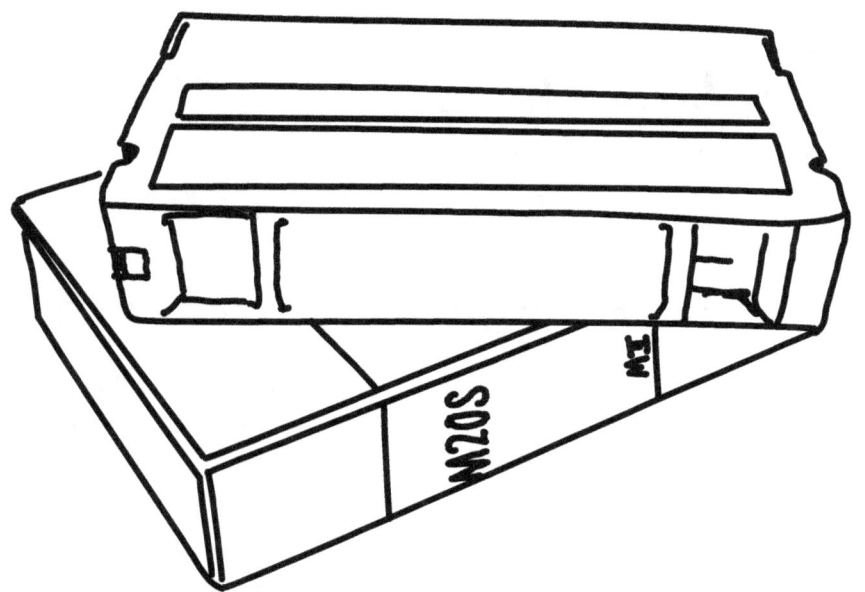

Fun fact
This format was an enhanced version of its predecessor, the failed M format; while the cassettes were similar, they used different magnetic tape formulas and signal processing methods

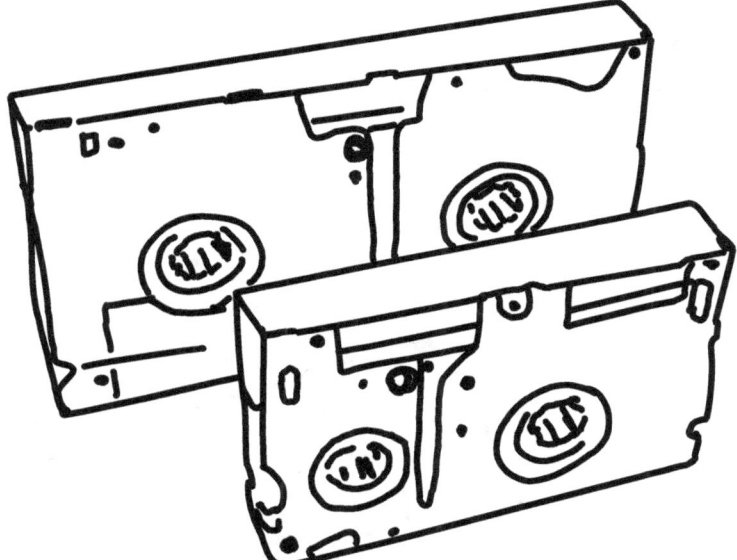

Fun fact
This format was developed to compete with Sony's Betacam SP format; while it gained some popularity, it largely lost out due to lack of reliable, affordable repair support

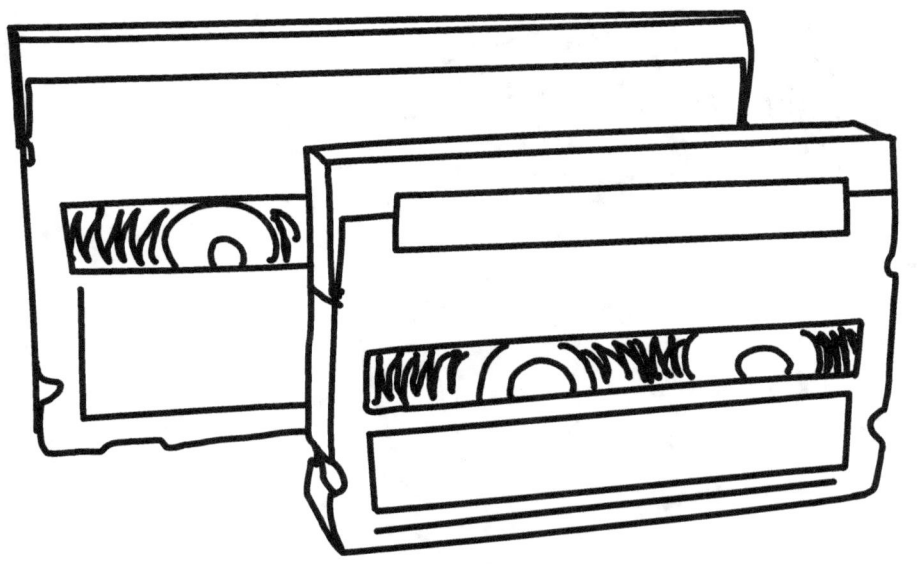

Fun fact
The M and MII formats, like the U in U-matic and the B in Beta, were named after the M-shape of the tape threading through a playback device (and VHS used this same technique)

Pixelvision

Format
analog

Developed by
Fisher-Price

AKA
**PXL2000, Sanwa Sanpix1000,
KiddieCorder, Georgia**

Capacity
11 minutes

Era
1987–1990s

Size
10 × 6.3 × 1.3 cm

Fun fact
**These camcorders
recorded low
resolution
audio and video
onto standard
music audio
cassettes**

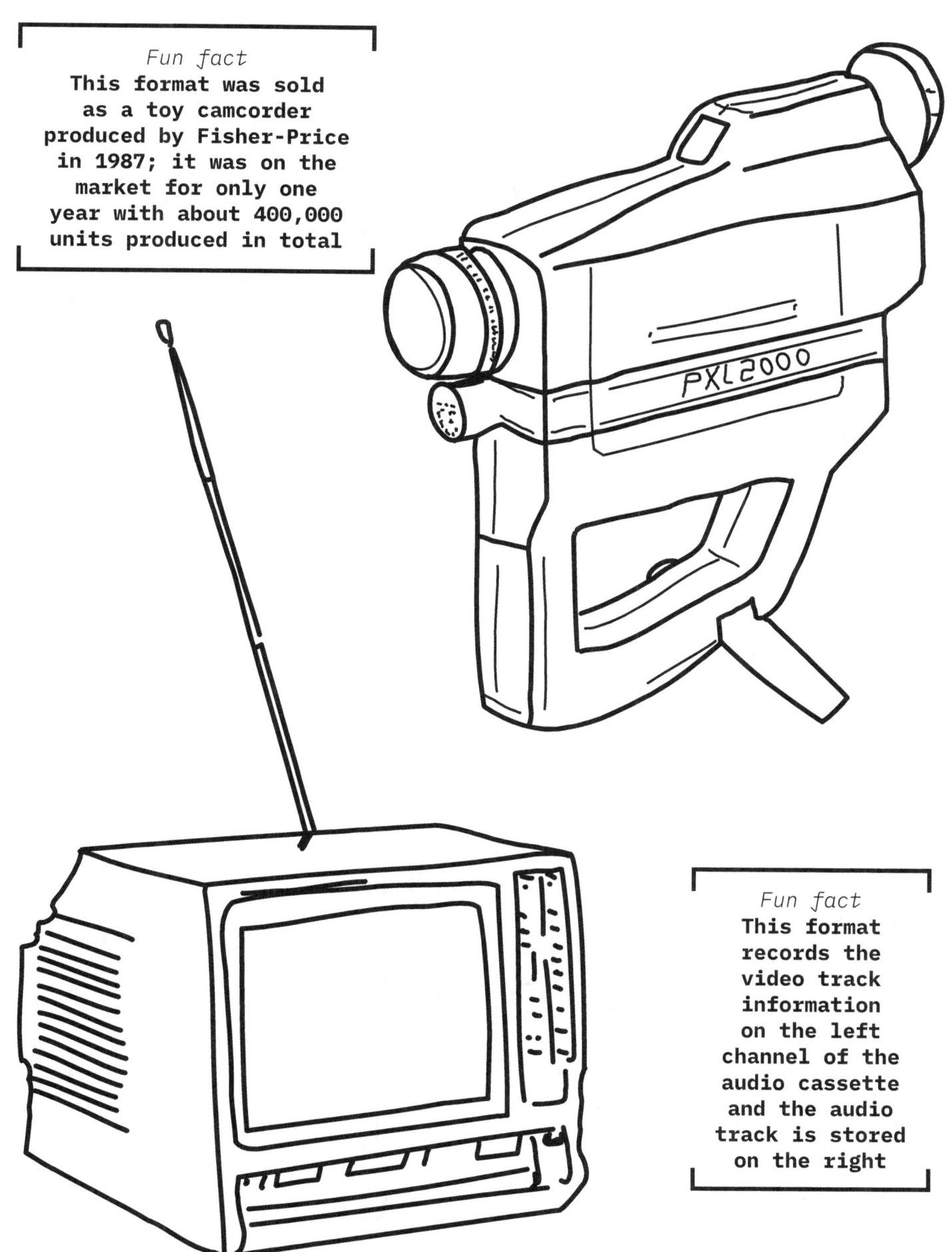

Fun fact
This format was sold as a toy camcorder produced by Fisher-Price in 1987; it was on the market for only one year with about 400,000 units produced in total

PXL2000

Fun fact
This format records the video track information on the left channel of the audio cassette and the audio track is stored on the right

D-2

Era
1988–2000s

Format
digital

Developed by
Ampex & Sony

Capacity
Small: 32 minutes
Medium: 94 minutes
Large: 208 minutes

Size
Small: 17.2 × 10.9 × 3.3 cm
Medium: 24.5 × 15 × 3.3 cm
Large: 36.6 × 20.6 × 3.3 cm

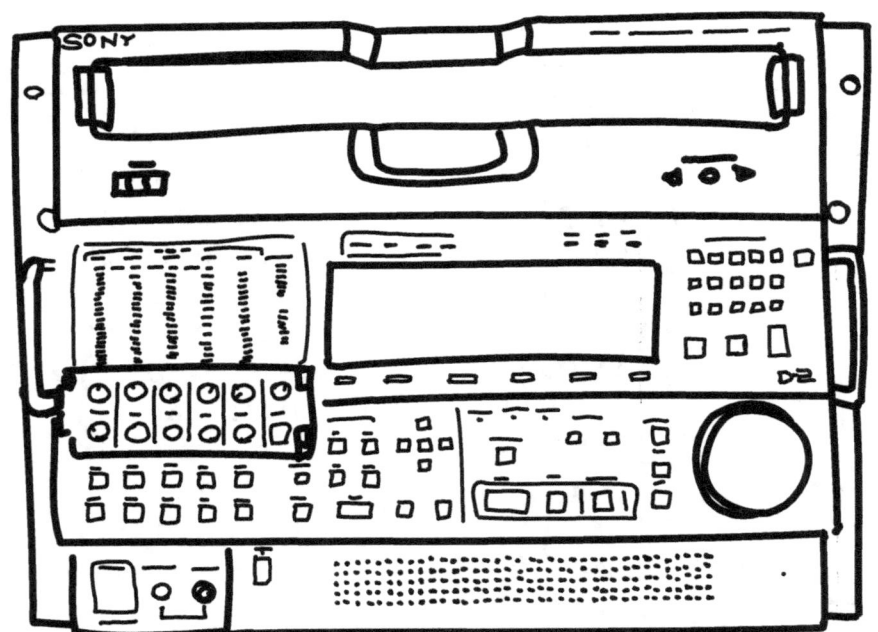

Fun fact
D2 machines are capable of using either serial digital video or analog video connectors

Fun fact
Ampex was awarded a technical Emmy in 1989 for the work on creating this format

Fun fact
While its predecessor, D1, used component video, D2 (and its successor, D3) used digital composite

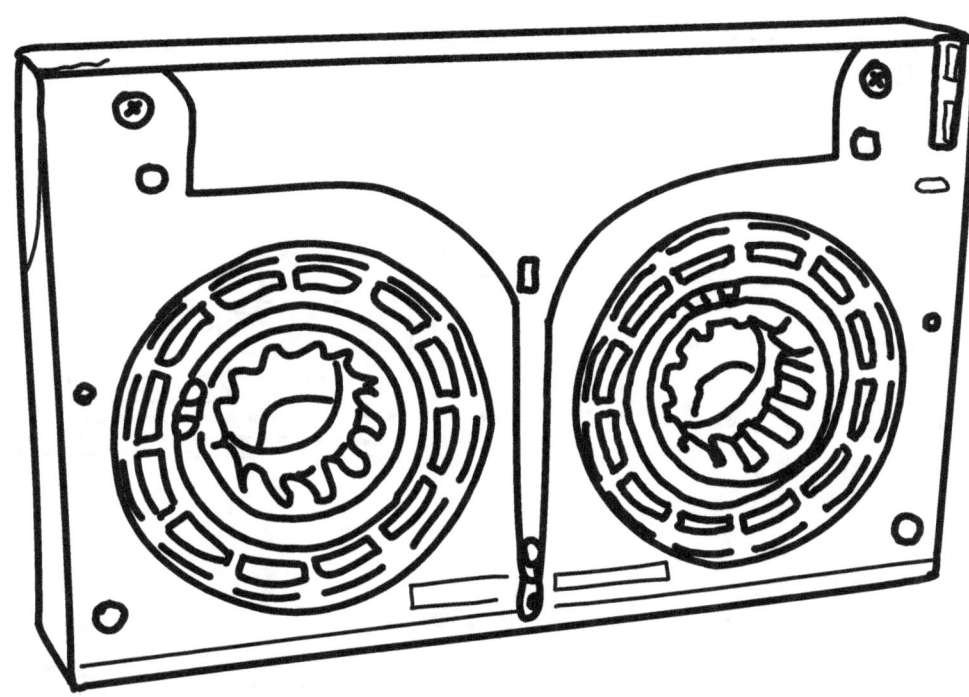

D-3

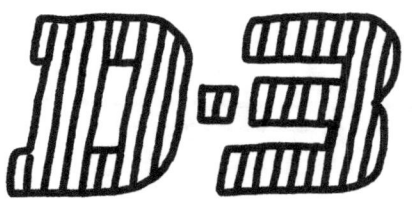

Format	
digital	

Developed by	Era	AKA
NHK	**1991–2000s**	**1/2" Digital**

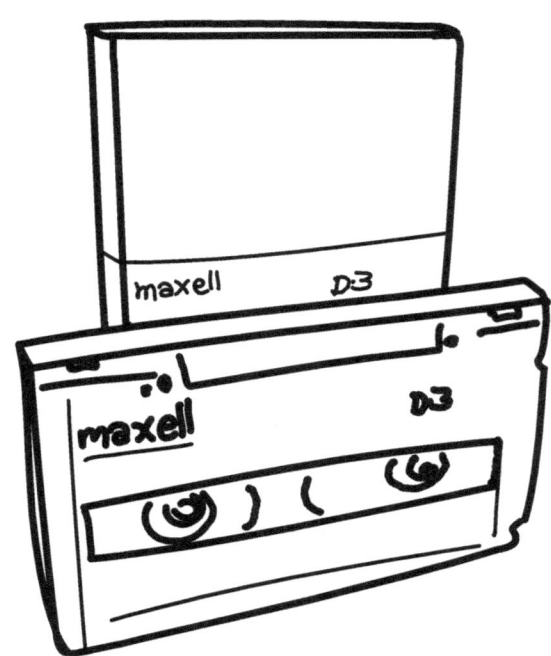

Capacity
Small: 50 minutes
Medium: 126 minutes
Large: 248 minutes

Size
Small: 16.1 × 9.6 × 2.5 cm
Medium: 21.2 × 12.4 × 2.5 cm
Large: 29.6 × 16.7 × 2.5 cm

Fun fact
The technology behind this format was developed by the NHK and commercially distributed by Panasonic

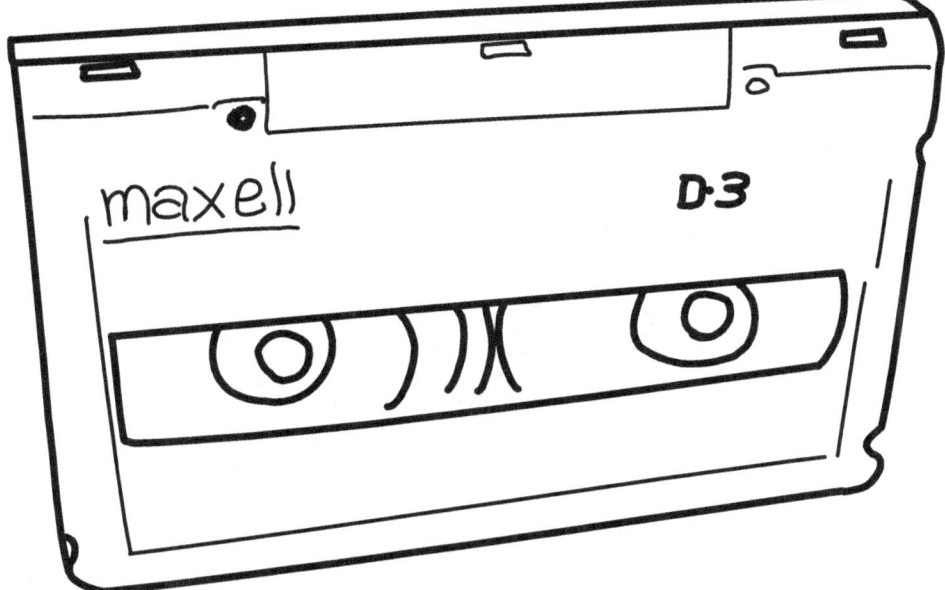

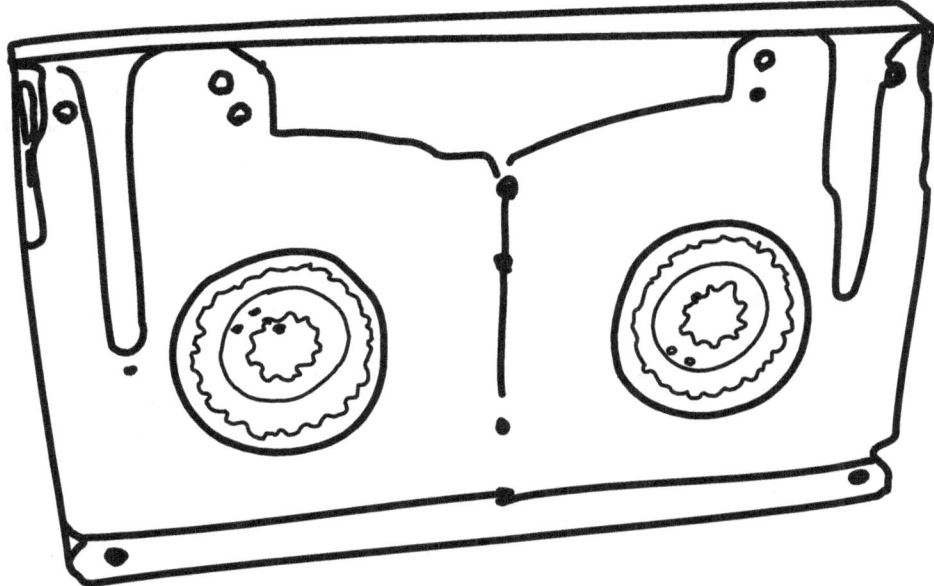

Digital Betacam

Format
digital

Era
1993–2016

Developed by
Sony

Capacity
Small: 40 minutes
Large: 124 minutes

Size
Small: 15.6 cm × 9.6 cm × 2.5 cm
Large: 25.4 cm × 14.5 cm × 2.5 cm

Fun fact
This format competed against the D1 format and was significantly cheaper

Fun fact
Digital Betacam decks included SDI coaxial digital connections, which allowed digital signals to be sent on existing coaxial wiring

Fun fact
These cases are usually blue and the tapes are various colors based on producer: Ampex (baby blue), Fuji (sea blue), and Sony (gray)

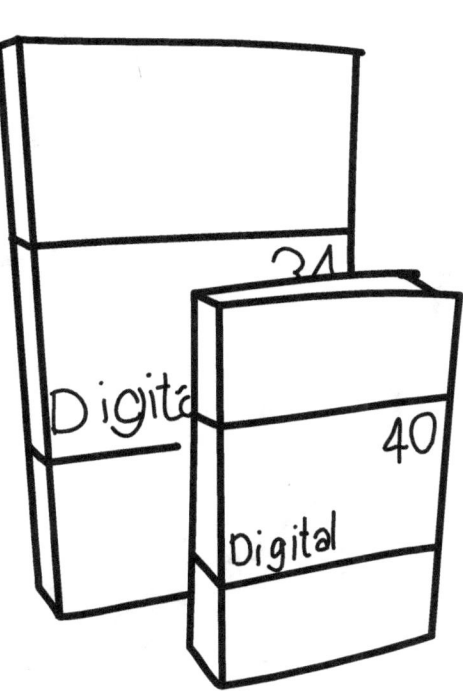

VCD

AKA
Video CD, Compact
Disc Digital Video

Format
digital

Developed by
Philips, Sony,
Panasonic, JVC

Size
4.7" discs

Capacity
74 minutes

Fun fact
This format had the most
commercial success in China
and Southeast Asia

Era
1993-2000s

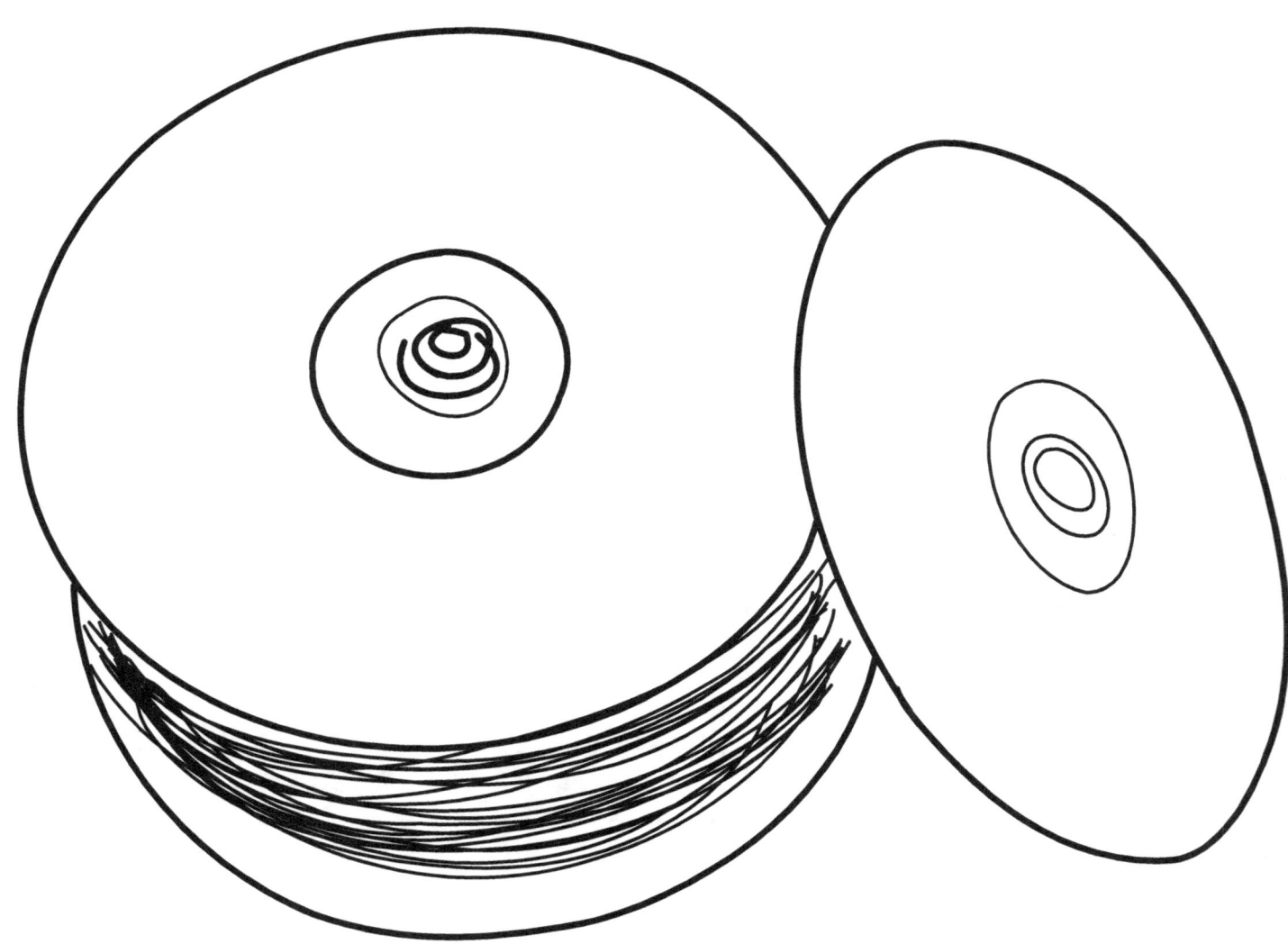

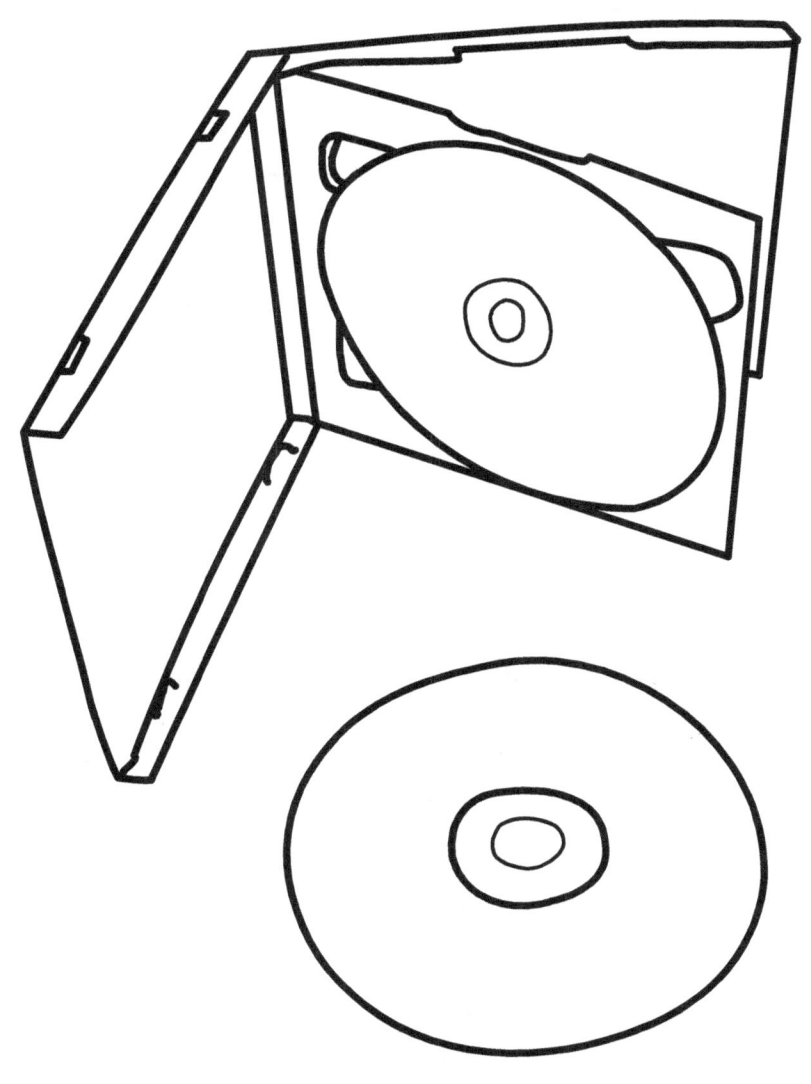

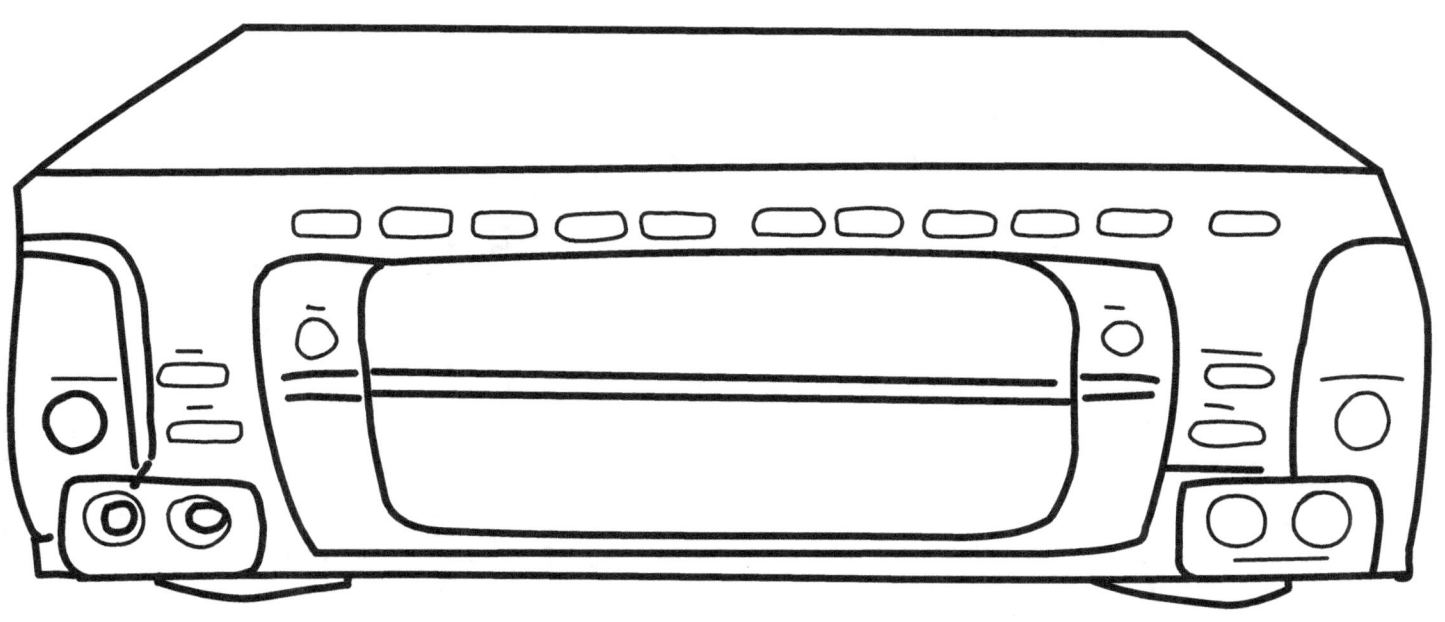

63

DV/MiniDV

Size
**Small: 6.5 × 4.8 × 1.2 cm
Large: 12.5 × 7.8 × 1.5 cm**

Format
digital

Developed by
Sony

Capacity
**Small: 66 minutes
MiniDV SP: 66 minutes
MiniDV LP: 132 minutes
Large: 126 minutes**

Era
1995–2010s

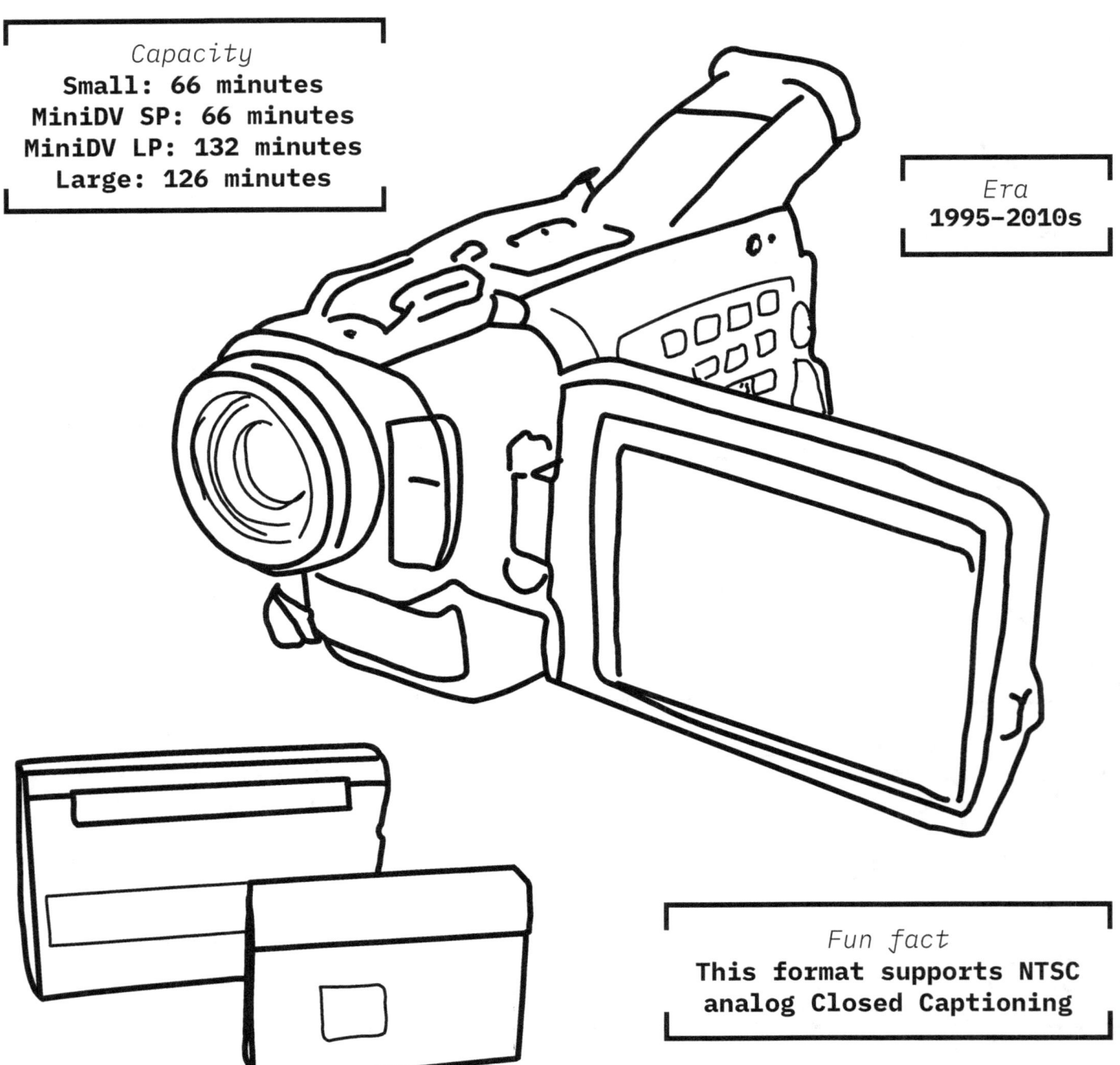

Fun fact
**This format supports NTSC
analog Closed Captioning**

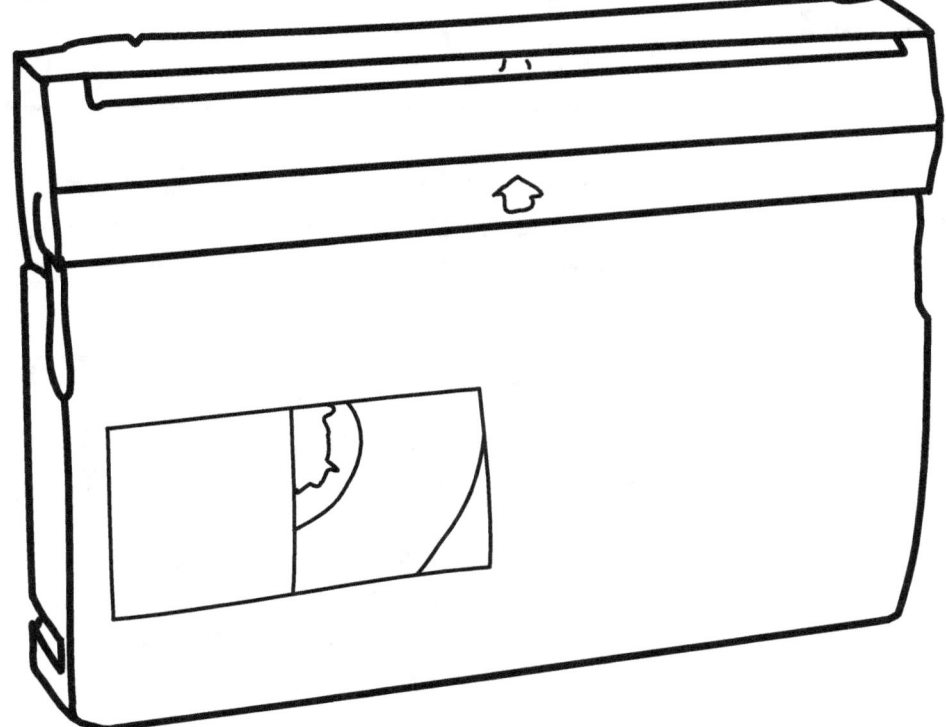

DVCPRO

Format
digital

Era
1995–2013

AKA
DVCPRO25, DVCPRO50,
D7, DVCPROHD, DVCPRO100

Size
Standard: 9.7 × 6.35 × 1.4 cm
Large: 12.4 × 7.6 × 1.4 cm

Capacity
DVCPRO25
Medium: 66 minutes
Large: 126 minutes

DVCPRO50
Large: 92 minutes
X-Large: 126 minutes

DVCPROHD
Medium: 33 minutes
Large: 64 minutes
X-Large: 126 minutes

Developed by
Panasonic

Fun fact
DVCPRO50, introduced
in 1997, used two DV
codecs in parallel,
doubling the data
rate over the original
DVCPRO to 50 Mbps

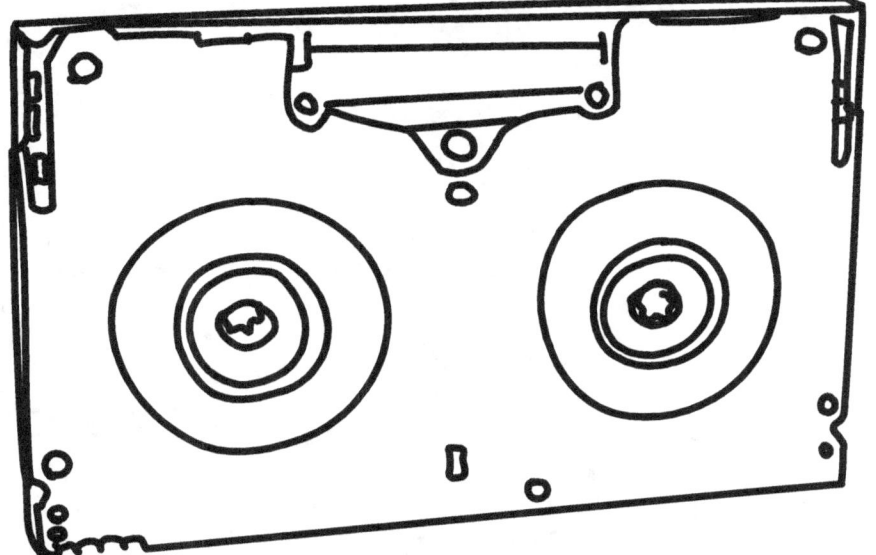

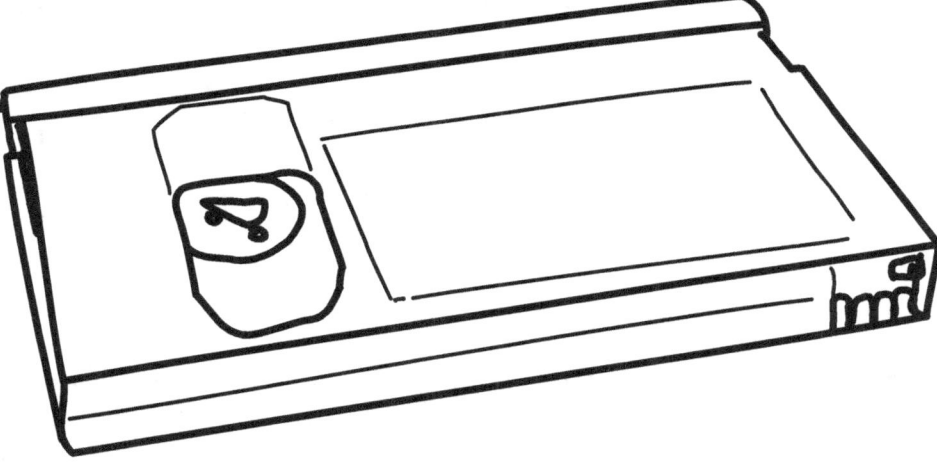

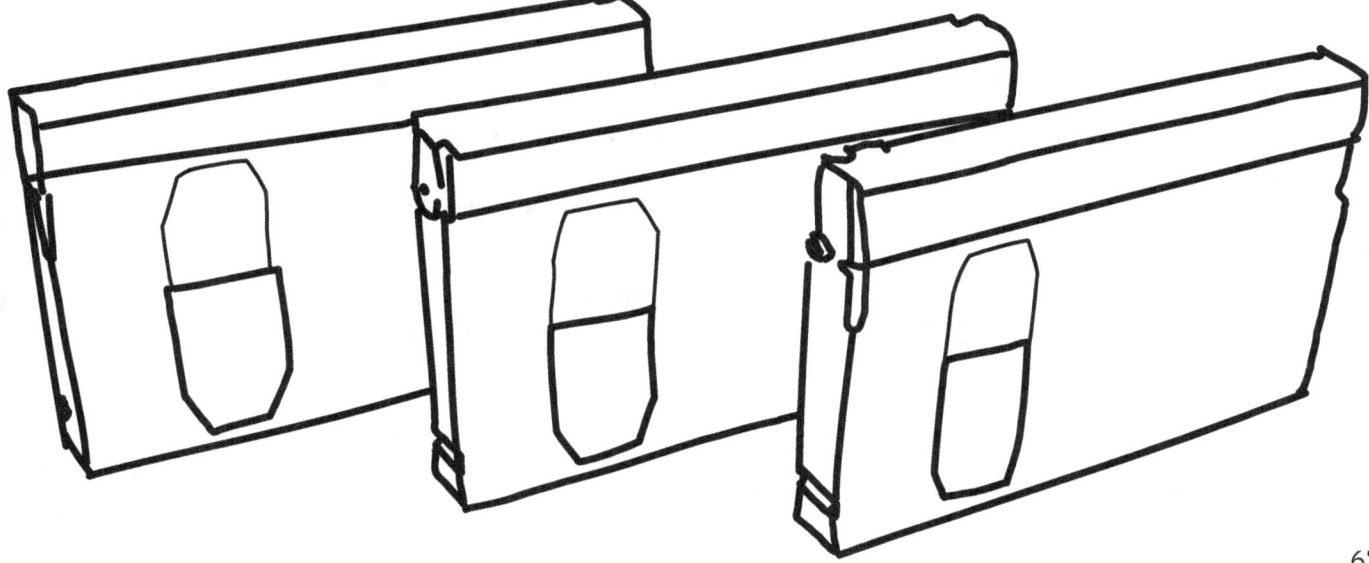

DVD

AKA
Digital Video Disc

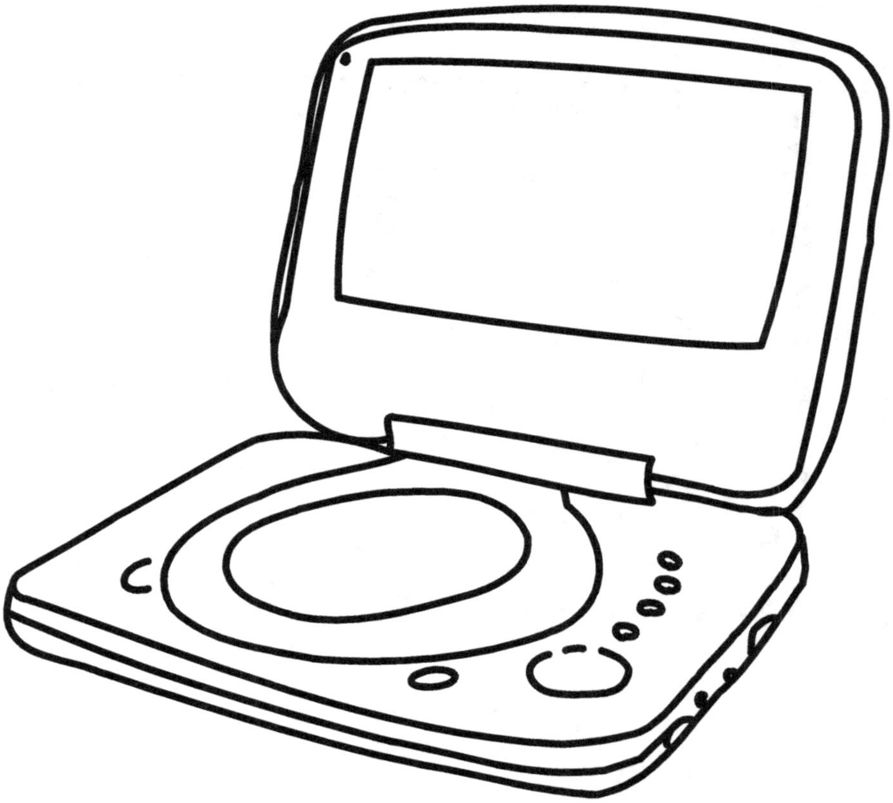

Capacity
4.7 GB (single-layer), 8.5 GB (dual-layer), up to 17 GB

Size
4.7" discs

Era
1996–2010s

Format
digital

Developed by
Sony, Panasonic, Philips, Toshiba

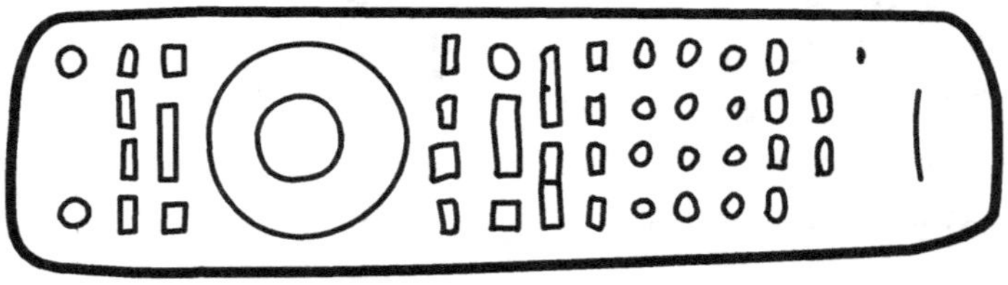

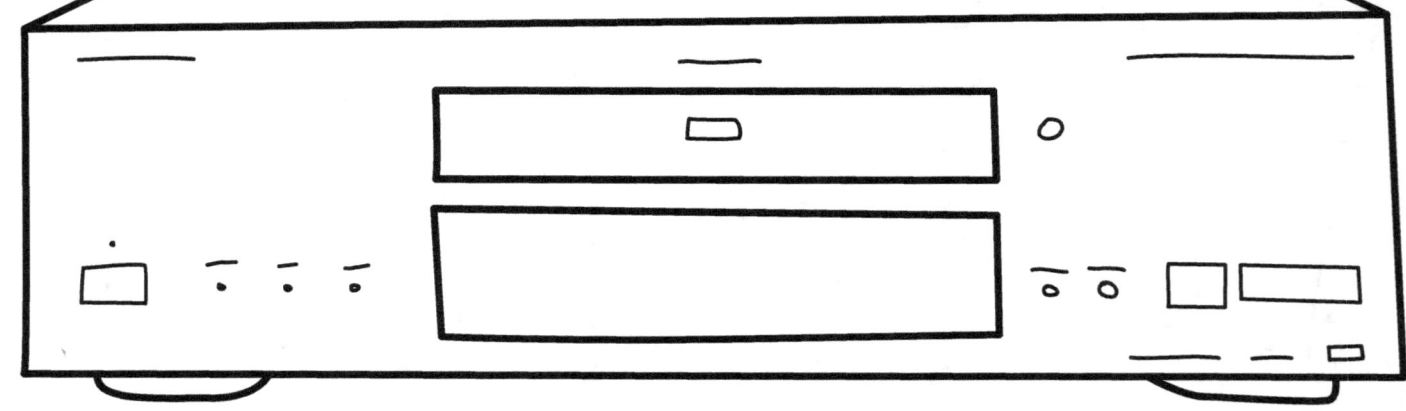

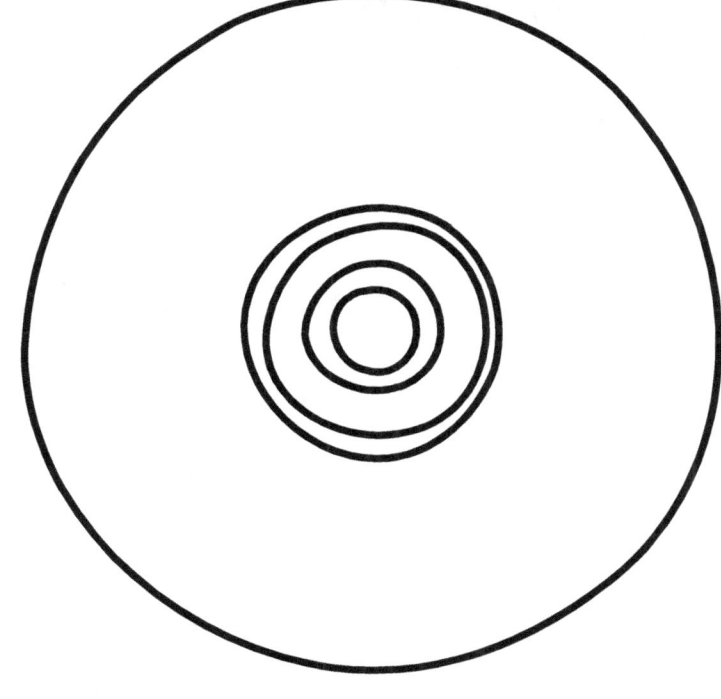

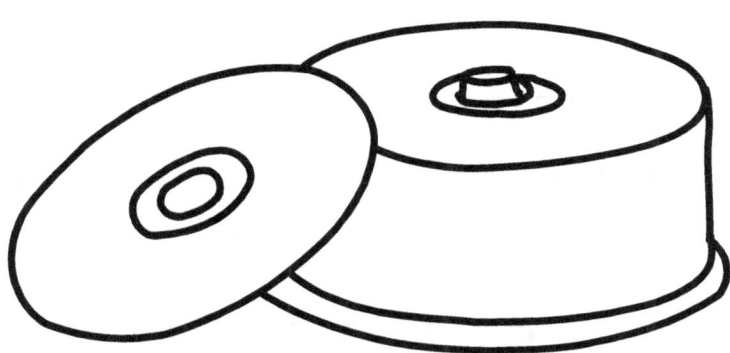

DVCAM

Format
digital

AKA
DV

Developed by
Sony

Capacity
Small: 40 minutes
Large: 184 minutes

Size
Small: 6.5 × 4.8 × 1.2 cm
Large: 12.5 × 7.8 × 1.5 cm

Era
1996–2010s

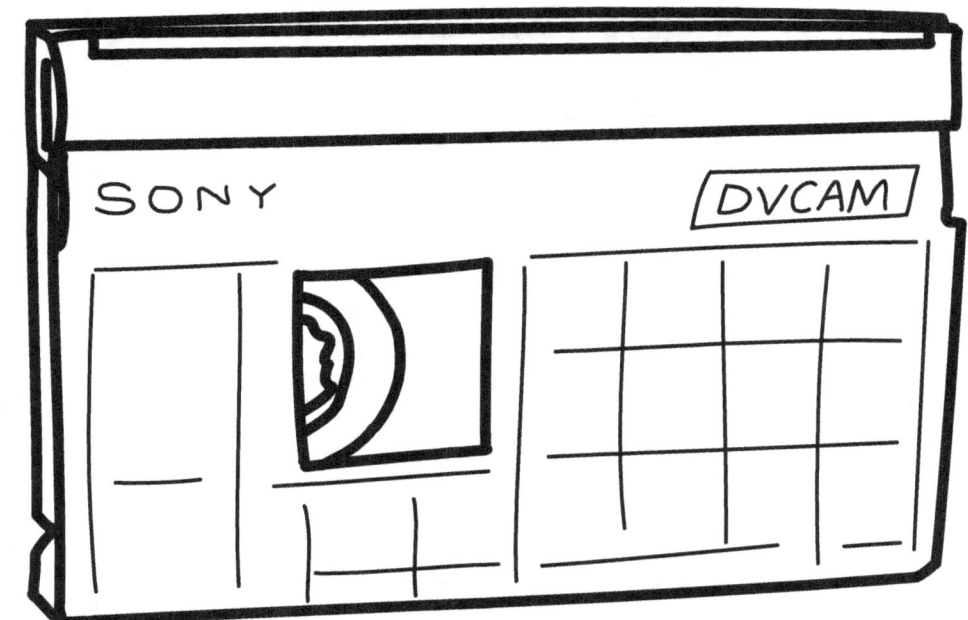

Fun fact
Many high-end DVCAM players would convert unlocked audio to locked audio on playback

Fun fact
This format was Sony's response to Panasonic's DVCPRO, both of which target the professional instead of consumer video market

Fun fact
This format has tracks that are 50% wider than standard DV, which increased robustness and reliability

HDCAM

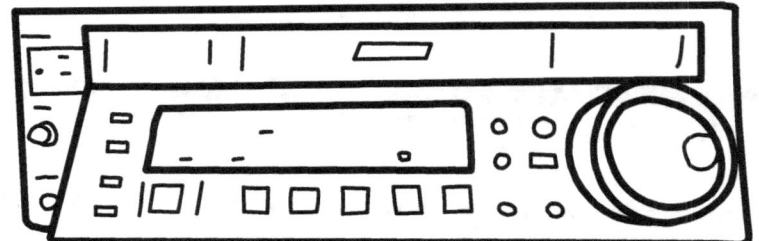

Era
1997–2016

Format
digital

AKA
SMPTE 367M, D-11

Capacity
Small: 40 minutes
Large: 124 minutes

Developed by
Sony

Fun fact
HDCAM cassettes were the same size as original Betacam but had a distinctive orange lid (and HDCAM SR cassettes were black with a cyan lid)

Size
Small: 15.6 cm × 9.6 cm × 2.5 cm
Large: 25.4 cm × 14.5 cm × 2.5 cm

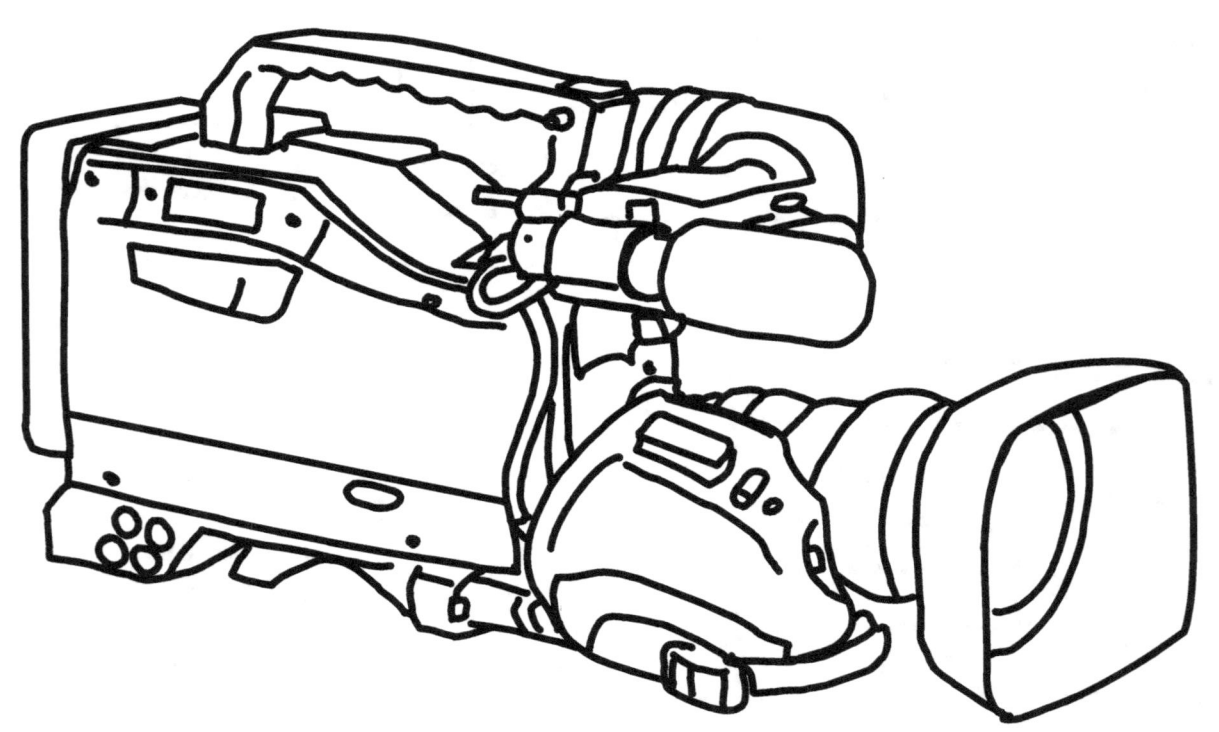

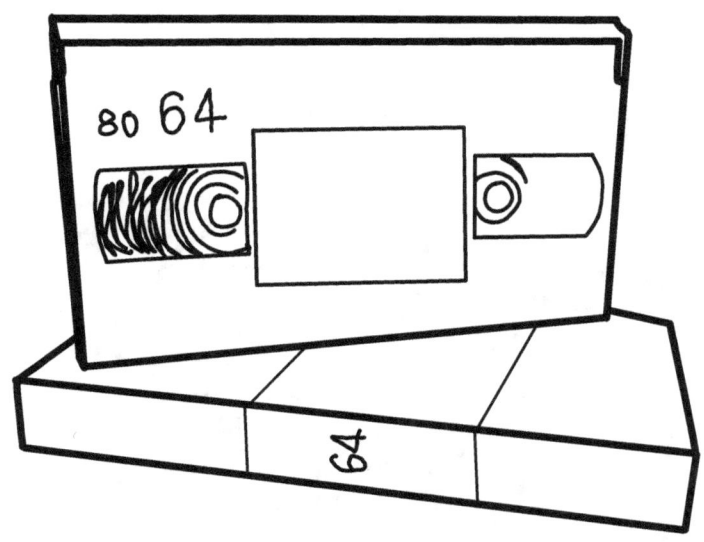

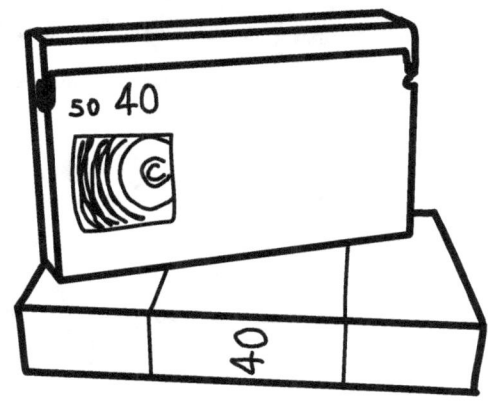

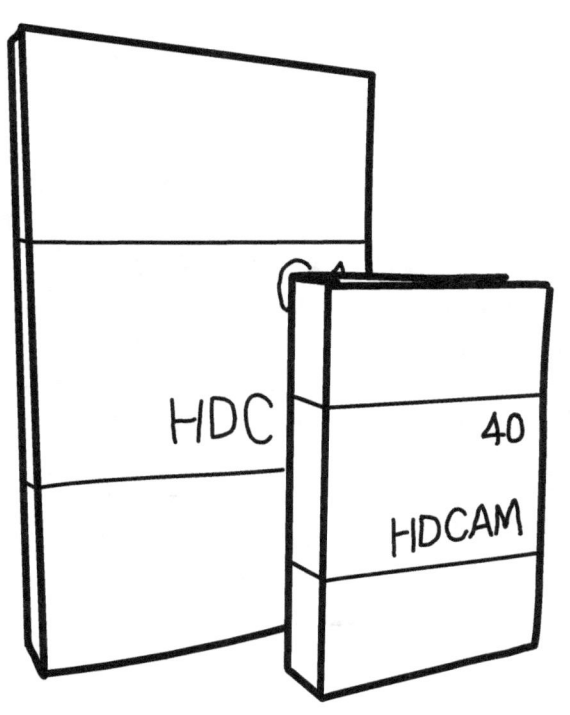

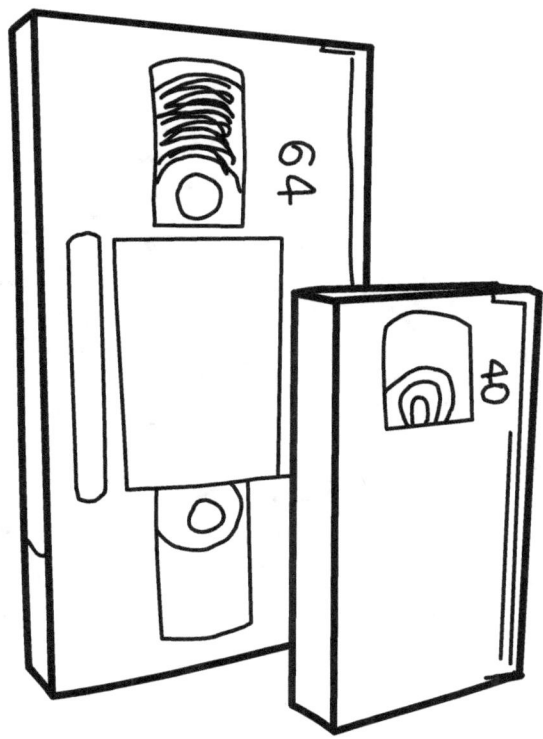

Digital8

Size
9.5 × 6.2 × 1.5 cm

Format
digital

Developed by
Sony

Capacity
90 minutes

Era
1999–2007

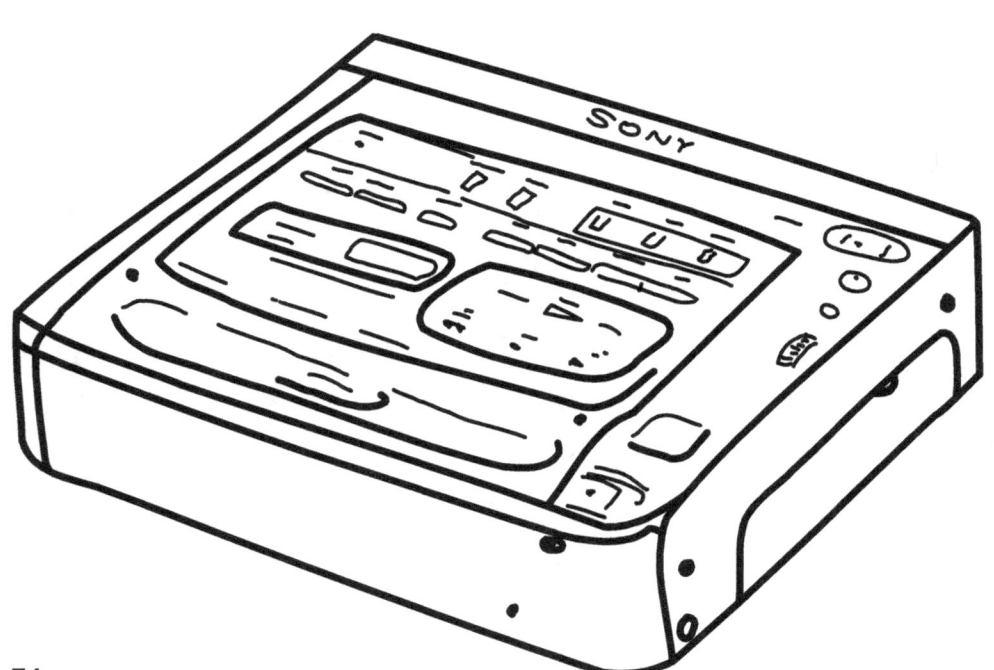

Fun fact
A Video8 or Hi8 tape can be recorded in digital mode, but a 2 hour tape could only store 1 hour of digital content

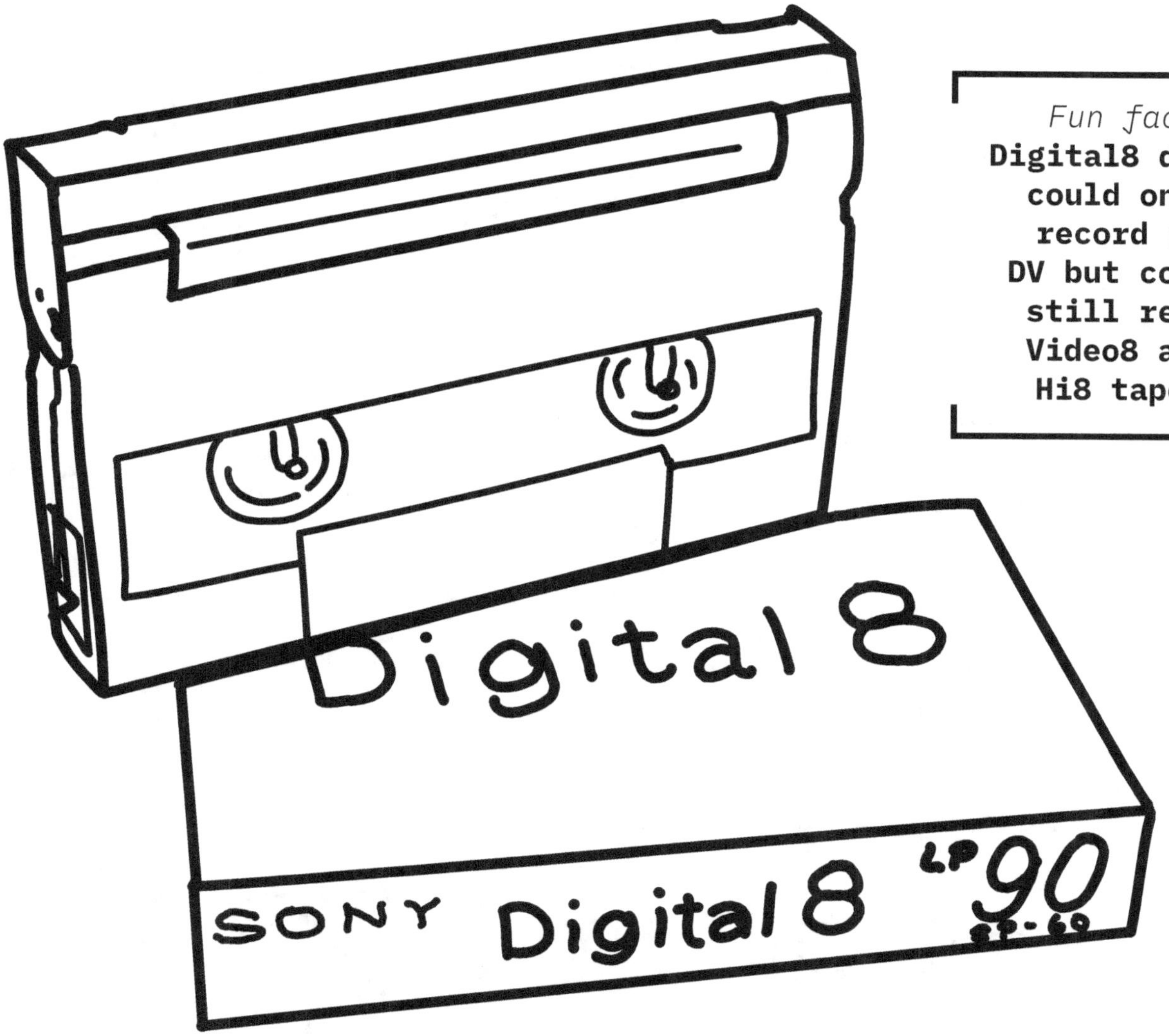

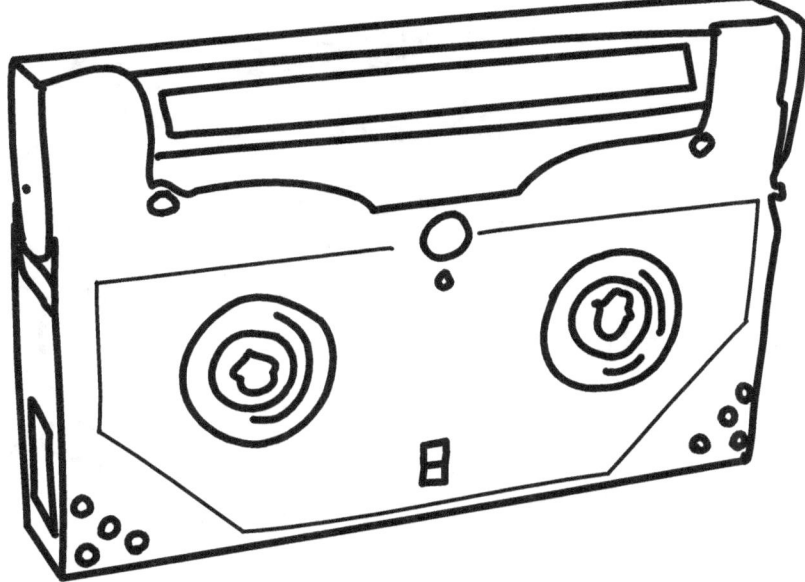

Fun fact
**Digital8 decks
could only
record in
DV but could
still read
Video8 and
Hi8 tapes**

Digital 8

SONY   Digital 8   LP 90 SP-60

Fun fact
**While predecessor
8mm formats
Video8 and Hi8
were analog,
Digital8 stored
digital video**

HDV

Format
digital

Fun fact
Two major versions of this format were HDV 720p (HDV1) and HDV 1080i (HDV2); the former was used by JVC and the latter was used by Sony & Canon

Era
2003–2011

Developed by
HDV Consortium

Capacity
MiniDV: 63 minutes
DVCAM: 276 minutes

AKA
High definition, HDV1

Size
MiniDV: 6.5 × 4.8 × 1.2 cm
DVCAM: 12.5 × 7.8 × 1.4 cm

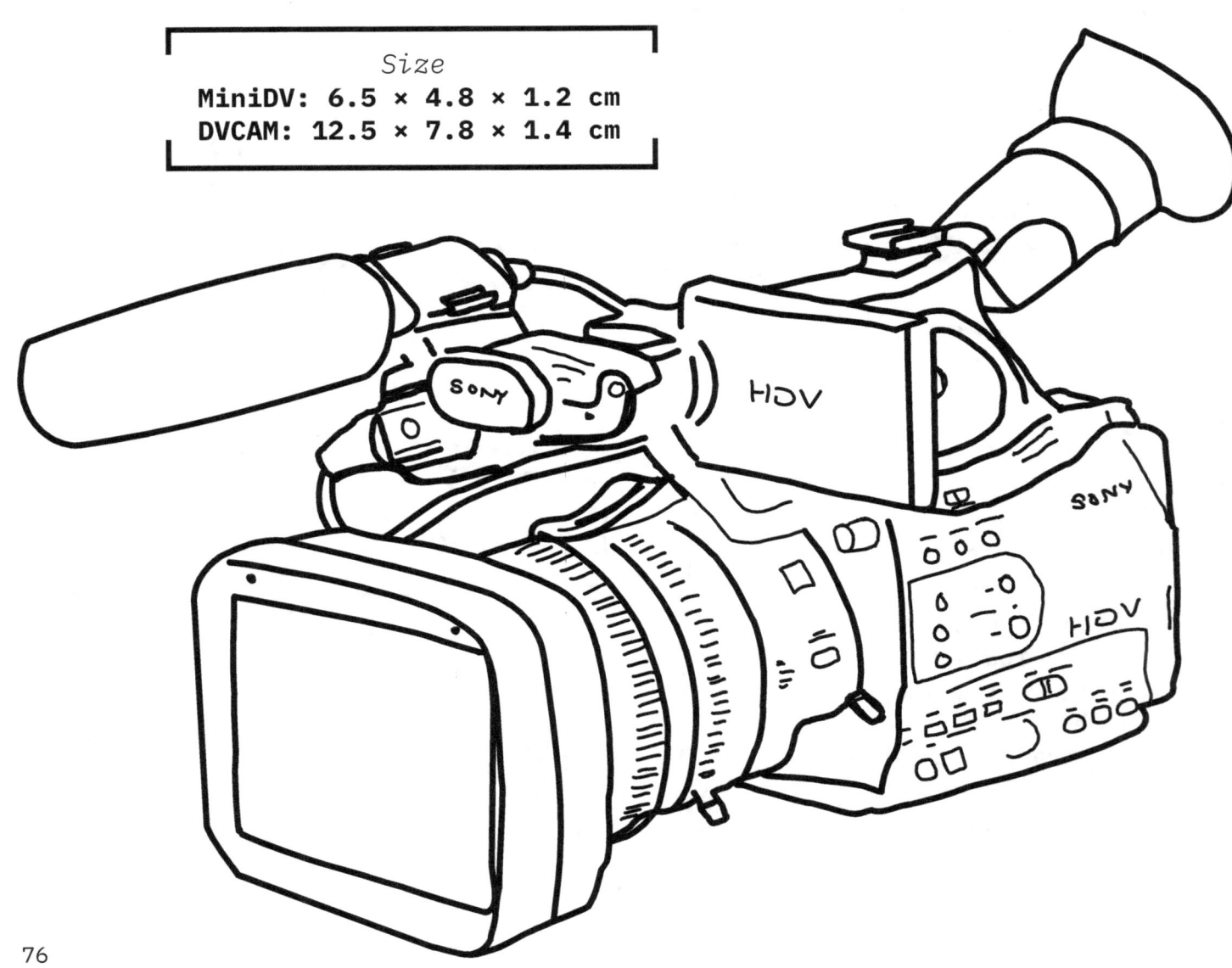

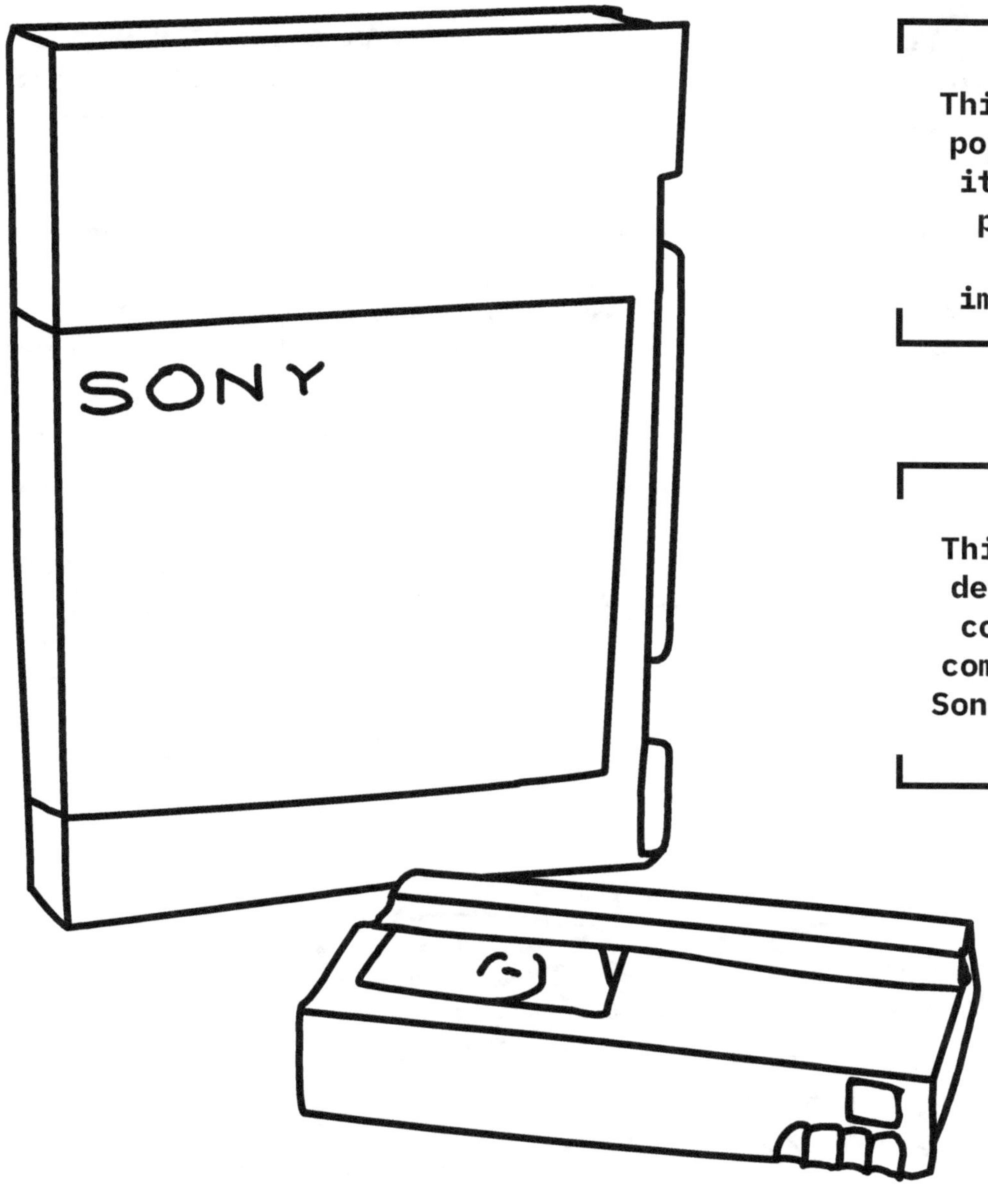

SONY

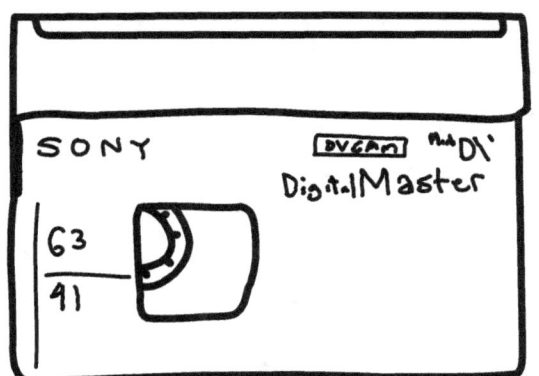

SONY DVCAM mini DV
Digital Master

63
41

XDCAM

Format
digital

Size
Various

Capacity
Various

Developed by
Sony

Era
2003–2010s

Fun fact
XDCAM uses random access solid-state memory media cards to store digital video

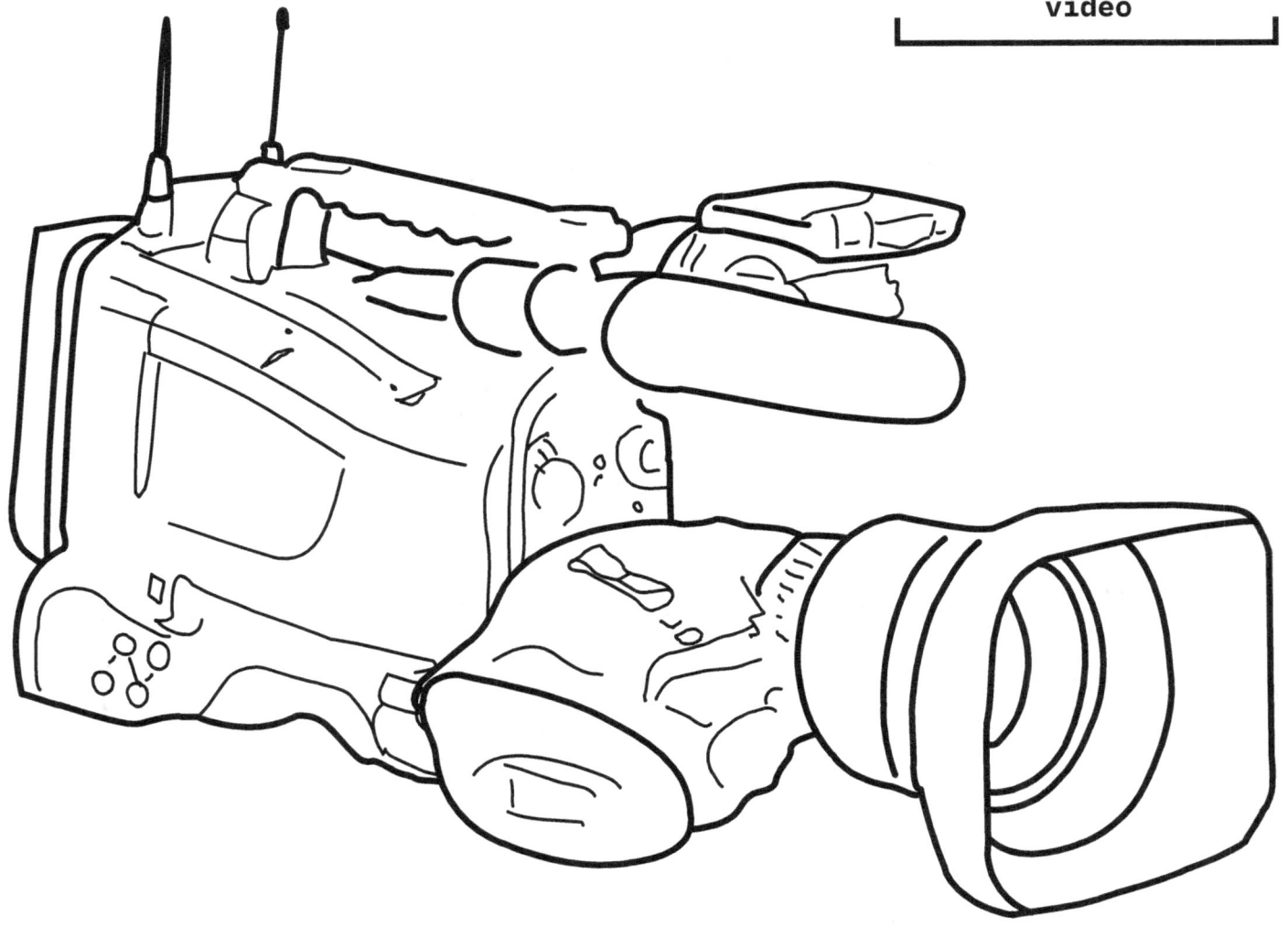

Blu-ray

Format
digital

Era
2006–2020s

Size
4.7" discs

Developed by
Blu-ray Disc Association

Capacity
**25 GB (single-layer),
50 GB (dual-layer),
up to 128 GB**

Fun fact
**While most discs are
region-free, the Blu-ray
Disc region coding scheme
divides the world into
three regions,
labeled A, B and C**

Fun fact
**These discs can have
up to four layers,
with each layer
offering more
potential capacity**

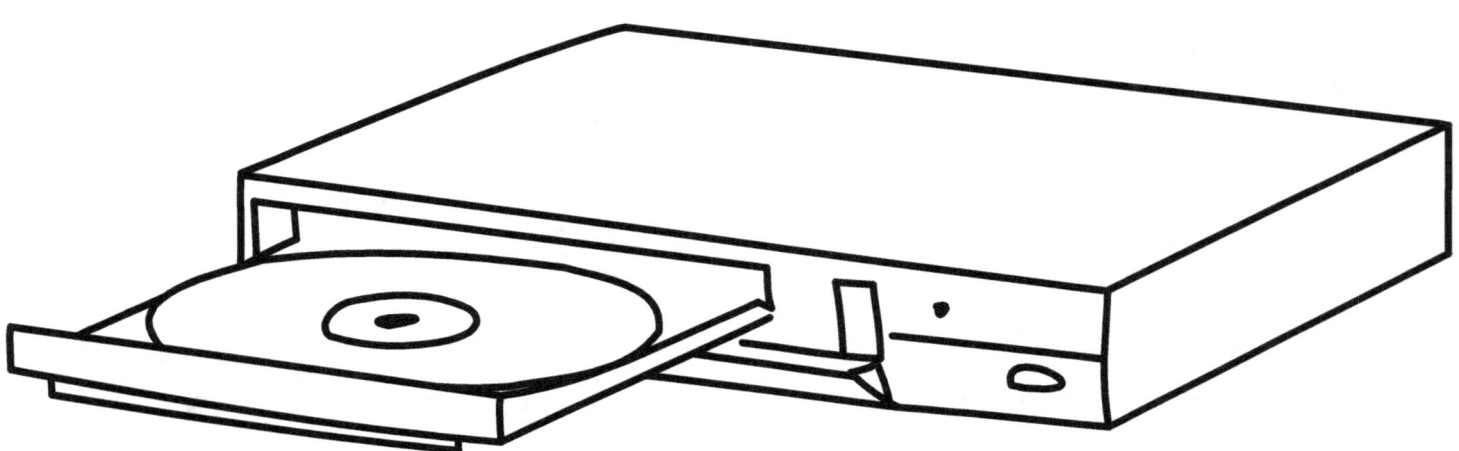

Fun fact
This format's name refers to the "blue" (violet) laser used to read the disc, which is stronger than the (longer wavelength) red laser used to make DVDs

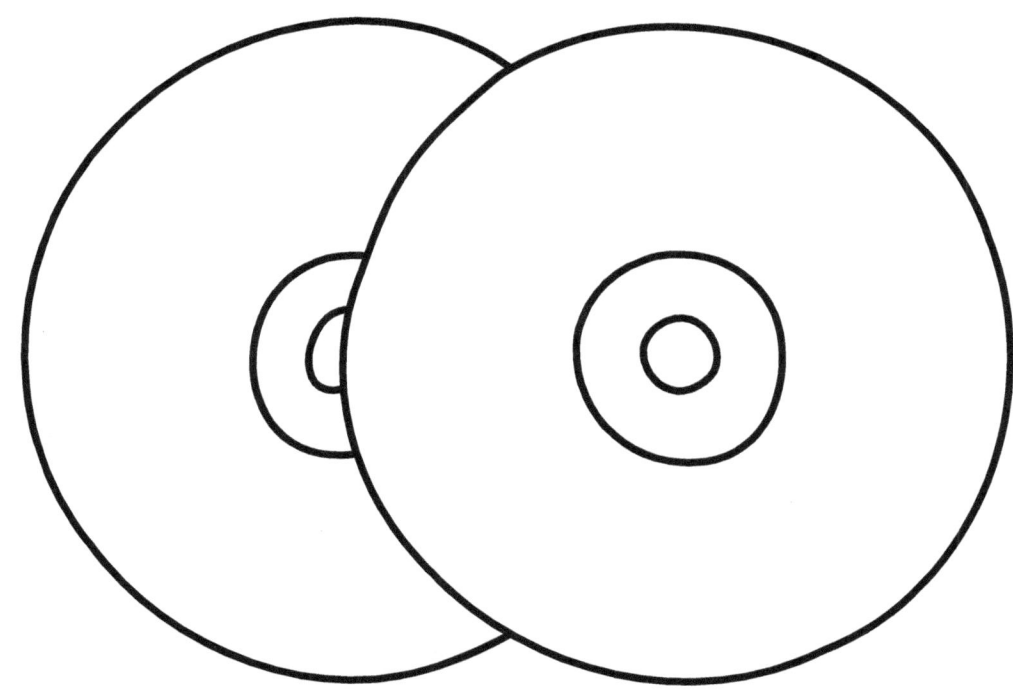

Blu-ray and DVD discs are the same size

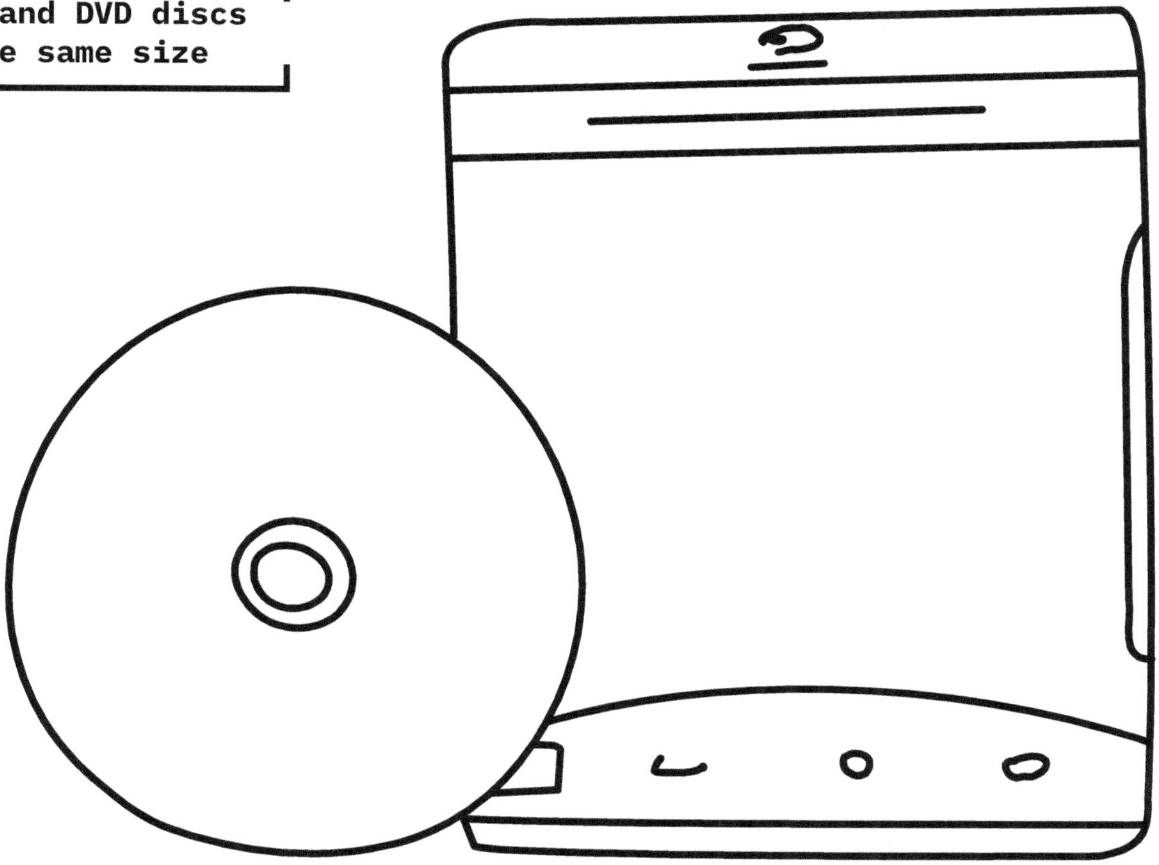

Acknowledgements

Thank you to my technical reviewers: Michael Henry Grant, Libby Hopfauf, Jackie Jay, and Ben Turkus. Your knowledge inspired me, your anecdotes warmed me, your enthusiasm motivated me, and your insights extend far beyond the pages of this book.

Thank you to my mom for telling me that the book looks great and that all the sentences make sense. I'm grateful to have inherited your curiosity and creative drive.

And of course thank you to Rory: for endless listening to my hopes and dreams, supporting every one of my ambitions, and for everything.

Also available

The Illustrated Guide to Audio Formats
The Illustrated Guide to Film Formats

About the Author

Ashley Blewer is an archivist,
educator, and software engineer
with over a decade of experience
building video analysis tools and
pipelines. Ashley specializes in
video and audio formats, digital
preservation, systems and
workflows, community support and
solutions, computer-to-human
interpretation, and teaching
technical concepts at all levels.

Learn more at
https://ashleyblewer.com